Jones' Instrument Technology Volume 5

Automatic Instruments and Measuring Systems

Rudolf Radnai

and

Edward G. Kingham

Butterworths

London · Boston · Durban · Singapore
Sydney · Toronto · Wellington

First published 1986

British Library Cataloguing in Publication Data

Jones, E. B. (Ernest Beachcroft)
Jones' instrument technology.—[4th ed.]
Vol. 5: Automatic instruments and measuring systems
1. Measuring instruments
I. Title II. Radnai, Rudolf III. Kingham, E. G.
IV. Jones, E. B. (Ernest Beachcroft).
Instrument technology
620′.0044 QC100.5

ISBN 0-408-01532-2

Library of Congress Cataloging in Publication Data

Jones, E. B. (Ernest Beachcroft)
Jones' Instrument technology.

Vol. 5 based on: Automatikus mérömüszerek és mérörendszerek/Rudolf Radnai.
Includes bibliographies and indexes.
Contents: v. 1. Mechanical measurements — v. 5. Automatic instruments and measuring systems/ Rudolf Radnai and Edward G. Kingham.
1. Engineering instruments. I. Noltingk, B. E.
II. Radnai, Rudolf. III. Title. IV. Title: Instrument technology.
TA165.J622 1985 620′.0028 84-4273

ISBN 0-408-01231-5 (pbk. : v. 1)

Photoset by Butterworths Litho Preparation Department
Printed Great Britain by Page Bros Ltd., Norwich, Norfolk.

Jones' Instrument Technology Volume 5

Automatic Instruments and Measuring Systems

Contents of other volumes

Contents

Preface

The purpose of this book is to encourage the use of automation in the field of electrical measurements. The enormous growth in science and technology and their application in many diverse areas has increased expectation of measuring systems to acquire more data at ever-increasing rates. The only practicable solution for such requirements is the use of automatic systems and these form a natural application for computer technology, providing optimum solutions in economy and technical performance. Nevertheless, the wide range of devices available creates its own problems and requires positive effort to resolve order from chaos. It is here that the international and national standardization organizations play a vital part.

In the book a brief review of a considerable number of standards is undertaken and particular consideration given to one that is receiving very wide support – the IEC 625 Interface System. This standard has received international acceptance by standardization bodies and companies involved in instrument development and manufacture.

For completeness, other relevant standards are also reviewed in modest detail. These include the main features of a number of processor backplane buses that are supported by one or more standardization bodies.

This work would not have been possible without the support of others, and the authors would like to acknowledge with grateful thanks the technical assistance of many instrument manufacturers, including Hewlett-Packard Corp, Tektronix Inc, E-H Research Laboratories, Rohde & Schwarz GmbH and Solartron, together with the dedicated efforts of the many committees and working groups of the various standardization organizations.

The foundation for the present work was laid in the original Hungarian book by my colleague, R. Radnai, whose extensive experience in instruments and measuring techniques at the Hungarian Academy of Science has made him an authority on the subject, whilst the high standard of work achieved by our translator, Mrs. B. Farkas, has proved invaluable in producing this new edition.

E. G. Kingham

Introduction

The main feature to emerge in the development of measuring technology is that of increasing automation. The need for automated measurement has become apparent in two main areas; in research and development, universal measuring systems with large computing capacity and data-processing facilities are required, whilst in the field of manufacturing quality control there is a growing need for automatic test equipment. This equipment must meet the immediate need but should be capable of being reconfigured for fresh requirements subsequently.

As the needs of the application areas have differed, so too have the development trends. In universal measuring systems there has been increasing standardization, whilst in the field of quality control instrumentation there has been a growing diversity, as each equipment supplier tends to go his own way. In this latter field the cost of equipment can also play a significant part in the choice of the system.

Owing to the diversty of the products and the rapid rate of their development, it is not practicable to carry out a detailed study of the individual products of every manufacturer. Accordingly, the book deals with general trends in electronic measuring automation, selecting specific examples only to illustrate principles. In particular, it majors on the design and use of systems based on the IEC Interface Recomendation (International Electrotechnical Commission Publication 625–1), otherwise commonly known as IEEE 488.

The book consists of seven parts.

After a short introductory Chapter, the interface system is discussed in Chapter 2, which details the connection of the individual units of an IEC-interface-based measuring system.

Chapter 3 is concerned with control units, computers, programmable calculators and special system controllers which are discussed in sufficient detail forit to be comprehensible for measuring technology experts interested in, but not necessarily well versed in, computing technology.

Chapter 4 deals with the building blocks of automatic measuring systems. In this Chapter the characteristics of program-controlled, 'intelligent' instruments are discussed in considerable detail according to their relevance in today's technology. A few modern instruments are described, emphasizing the salient features relevant to their use in automated systems.

In Chapter 5 the design and assembly of automatic systems are discussed. The main theme of the discussion is again the IEC interface, but the hardware and software design descriptions are of a general nature, and therefore valid to any automated system.

Chapter 6 deals with IEC system analysers, with their merits and uses. In this Chapter the reader is acquainted with some of the problems encountered when an automatic measuring system is set up and operated.

Chapter 7 reviews alternative serial and parallel systems and interfaces and the standards currently available. Application of these will lie mainly in the research and development field but there is little doubt that industry will make growing use of such standards. Sufficient information and references are therefore provided to permit a valued judgement to be made of their relevance to potential schemes.

1 Automation of measuring techniques

1.1 Principles of automatic measurements

Measurement is an activity in which a number quantifies a physical entity, determined by comparison to an accepted unit. The physical entities may consist of either analogue or discrete quantities, constant or time variable. Most physical quantities are analogue and time variable.

The measuring instrument receives the input of the quantity to be measured and converts this to a numerical value of the output signal. Measuring instruments can be classified in various ways. One system classes instruments as directly or indirectly measuring. Direct measurements can be used for those physical quantities that cause a directional pointer deflection or some other similar feature. Non-directional quantities can be measured only by indirect methods and hence conversions will be necessary before comparisons are made during the measuring process.

According to another classification system, there are analogue and digital instruments. Analogue instruments proportionally convert the quantity to be measured into another quantity, which generally results in a pointer deflection which is then converted to a numeric value by means of a scale. This 'quantifying factor' indicates the relationship of the measured quantity to the chosen unit. In theory, the set of analogue values available is infinite, because the deflection of the pointer is the function of the measured quantity. In practice, however, owing to limitations in subdividing and reading the instrument scale, analogue instruments are significantly less accurate than digital instruments.

The main characteristic of the digital instruments is 'quantizing', where the continuously-changing values of the analogue quantity are converted into a limited set of discrete values. Hence the accuracy of digital instruments is limited by the quantizing error. In practice, however, the resolution of the digital instruments is considerably better than that of the analogue instruments. Furthermore, the numerical display facilitates fast reading of the results and eliminates subjective errors made by the observer.

An important advantage of digital instruments is that they can be incorporated in automatic measuring systems suitable for executing complex measuring tasks.

1.1.1 Measuring systems

A measuring system is an assembly of instruments and auxiliary units devised to carry out a certain measurement task.

Figure 1.1(a) shows a general measuring system for manual operation. Its components are: the signal

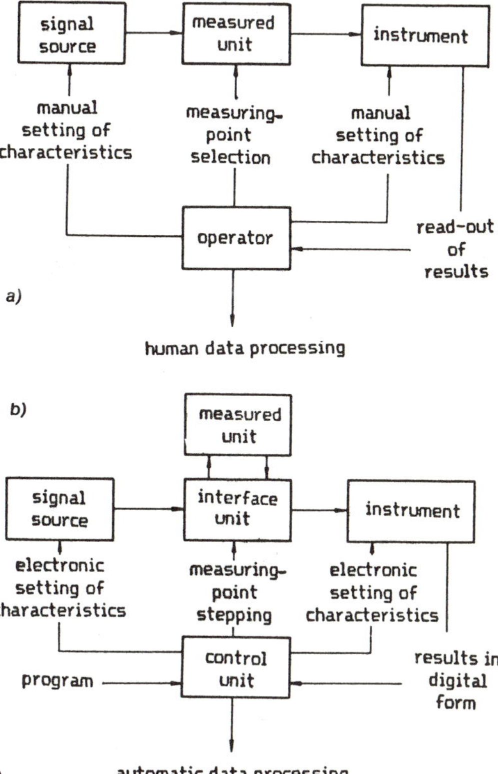

Figure 1.1 Structure of measuring systems. (a) Manual operation, (b) automatic control

source (e.g. a generator), which provides the input signals necessary for the measurement, and the instrument (e.g. a voltmeter), which senses the output signals. These two units are directly connected to the unit to be measured. With traditional, manually-controlled measurements there is also an operator present whose tasks are:

(a) selecting the output characteristics of the signal source,
(b) selecting the input characteristics of the instrument,
(c) preparing the unit for the measurement (e.g. provision of power supply, selection of measuring points, etc),
(d) reading the measurements, and
(e) evaluating the results.

In automatic measuring systems (Figure 1.1(b)) the control unit takes the place of the operator in controlling the signal source and the instrument, measurement step determination and data storage and processing.

In automatic measuring systems the signal sources and instruments are controlled by digital signals and the output is in digital form. The signal source and the instrument are connected to the unit under test via cables and an interfacing unit; this represents another important difference from the manually-controlled system.

1.1.2 Information transfer

The operation of automatic measuring systems is based upon information transfer between its units. The information traffic between the individual units therefore serves two purposes:

(a) control, and
(b) data transfer.

One of the most important parts of measurement is the setting-up of the instrument characteristics according to the measured quantity:

- for a *signal source type instrument* (e.g. function generator) this consists of the selection of the output signal form (e.g. impulse signal, saw-tooth signal, sine-wave) and of the setting-up of the signal characteristics (e.g. frequency, amplitude, packing density);
- for a *measuring type instrument* (e.g. multimeter) it will consist of the selection of the operation mode (voltage, current or resistance measurement) and the setting-up of the measuring limits.

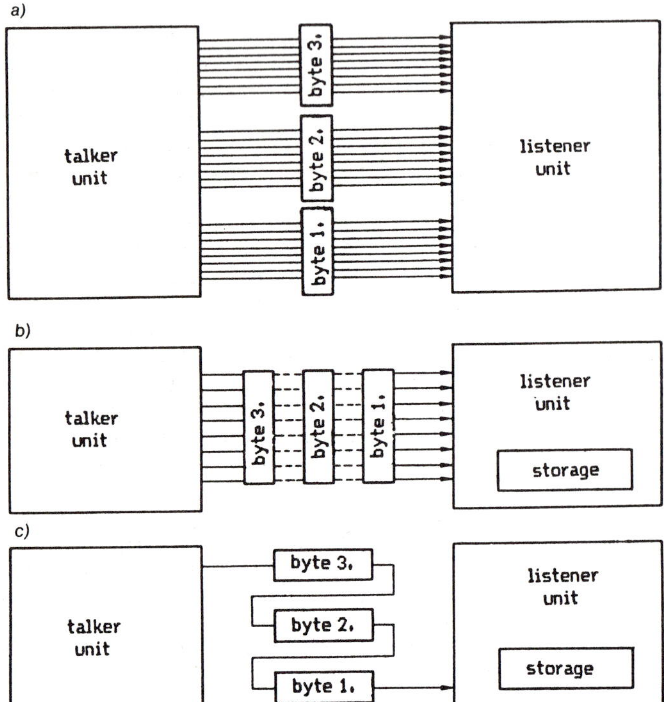

Figure 1.2 Digital data transfer methods. (a) Parallel, (b) byte-by-byte, (c) bit-by-bit

When traditional, manually-controlled instruments are used, the above operations are carried out by manipulation of the controls on the instrument panel. The control of programmable instruments is executed by digital command signals initiating predetermined logical state functions. In automatic measuring systems both the measurement results and the control signals are transferred in digital form. According to the task of the measuring system, it is possible to carry out different operations on the measurement results produced by the instrument. In many cases it is sufficient to compare the measured value with a pre-set limit to obtain a simple go/no-go decision. Similarly, another task could be the registration and storage in memory of the measured values in preparation for further digital processing.

With more complex, computer-controlled systems the control of the instruments or the sequence of the measuring points can be selected according to the measured values.

The form of the information transfer between the individual units is particularly important. Figure 1.2 shows the three basic digital information transfer methods. Each has its advantages and disadvantages, as indicated below.

With the *parallel method* (Figure 1.2(a)), all signals are transmitted simultaneously. The only advantage of this method is speed, whilst a penalty is paid in the multiplicity of cable and connector requirements.

With the *byte-by-byte method* the information is transferred sequentially, in 8-bit bytes (Figure 1.2(b)). This method represents a compromise between purely parallel and purely serial transfers in respect of cost and speed. A significant advantage is gained in that certain units of automatic instrumentation systems (e.g. punched-tape readers, calculators) operate on byte-by-byte data traffic and consequently their interfacing is easier when the external data traffic is similarly organized.

Figure 1.2(c) shows the *bit-by-bit (serial) data transfer* arrangement, where a single signal line is all that is needed, as the information transfer takes place bit-by-bit, sequentially. This method is comparatively slow and therefore, despite its simplicity and cheapness, it is normally used only to connect devices that are themselves slow in operation. In the cases of byte-by-byte and bit-by-bit transfers, the need for a listener buffer storage unit is an important factor when making comparisons between the digital information transfer methods. As far as the data traffic within the automatic measuring system is concerned, this means that the operation of the interface units will become more complicated.

The amount of information flow in control and data transfer can vary greatly, according to the instruments used and the characteristics of the measurements. For example, the digital control equivalent to a six-position switch is 3 bits, and three signal cables are necessary. For a nine-character digital counter the transfer of 36 information bits is necessary, which requires 36 signal lines if parallel transfer is used. This large number of signal lines would make the construction of automatic systems very expensive and complicated, and therefore purely parallel information transfer is normally used only if a high information transfer speed is necessary.

1.2 Classification of measuring systems

There are several ways of classifying automatic measuring systems. One relates to the characteristics of the measurement to be undertaken with the system. Some instrumentation systems are suitable for general purposes and some are custom-made for specific applications.

General purpose instrumentation systems are mainly used in research and development. The individual units of these systems are not necessarily produced by the same manufacturer and can often be used on their own. The units of such instrumentation systems – especially the measuring instruments – can be changed according to the requirements of the user.

Whilst the units of custom-made measuring and testing apparatus are usually produced by the same manufacturer, the controlling computer is often an exception to this and even the automatic system manufacturers obtain these from the computer suppliers. In general, the units of a custom-made measuring system cannot be used individually and usually the system configuration cannot be altered easily. These systems are used mainly in industrial production areas, where fast and reliable operation is more important than great accuracy.

Automatic measuring systems can also be classified according to the interconnection method. The star or radial system, shown in Figure 1.3(a) has all its units connected via individual signal cables to the control unit. This is the simplest connection method. All early automatic instrumentation systems were made this way. Simplicity is the key advantage of this connection method, but any change or expansion of the complete system is frequently very difficult, requiring complete redesign of the control unit. A further disadvantage is that the two-way data traffic between the control unit and each instrument requires a large number of multiwire cables, thus reducing the reliability and ease of handling of the system.

a)

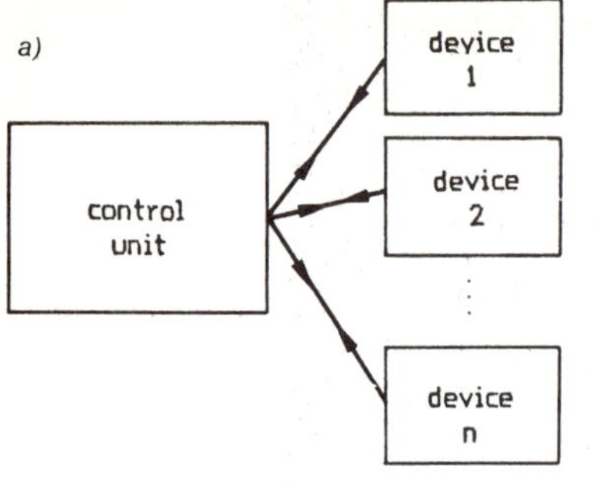

b)

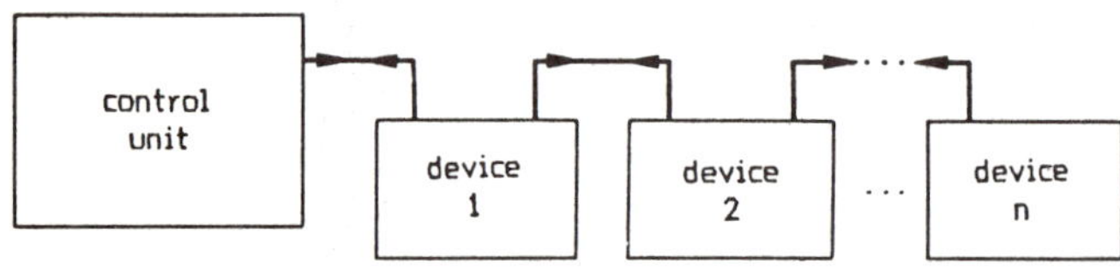

c)

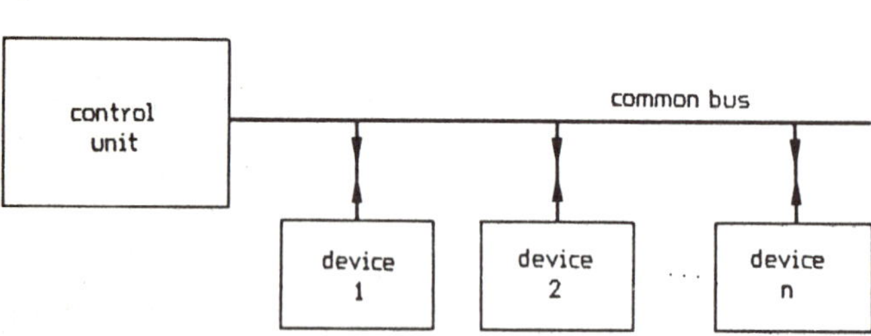

Figure 1.3 Organization methods for automatic measuring systems. (a) Radial organization, (b) serial organization, (c) bus organization

Figure 1.3(b) shows the block diagram of the point-to-point or daisy-chain system used for the interconnection of the individual instruments. An important characteristic of this system is that the simple serial connection ensures the hardware priority internally.

The most up-to-date interconnection method is the bus-line or party-line system, shown in Figure 1.3(c), in which a single signal conducting system (i.e. the common bus) connects the units of the system together. The standardized instrument interface system based upon the IEC recommendation (described in detail later) is an example of this. The advantage of the bus-line system is the minimal amount of connecting cables required. Also, the system can be expanded or modified according to the needs of the user and the fullest flexibility of control is possible.

Another classification of automatic instrumentation systems is possible with respect to the control mode used. Externally-controlled instrumentation systems are programmed by either some external peripheral devices (e.g. a punched paper tape reader), or by a fixed-program sequential control unit in the system. A common feature of these systems is that the programming of the controller is fixed and not readily altered. The control of the measuring system units is executed by the control unit, using commands stored on punched paper tape, magnetic tape or semiconductor storage; hence the sequence of the commands is fixed. These systems are comparatively cheap and there is no need for software experts in constructing and operating them. However, a considerable disadvantage is that they are less adaptable and generally operate rather more slowly than corresponding computer based systems.

Figure 1.4 shows an externally-programmed instrumentation system consisting of a Type CP 70A control unit by California Instruments and two digital multimeters connected to it. These can be programmed in Read Only Memory (ROM) to

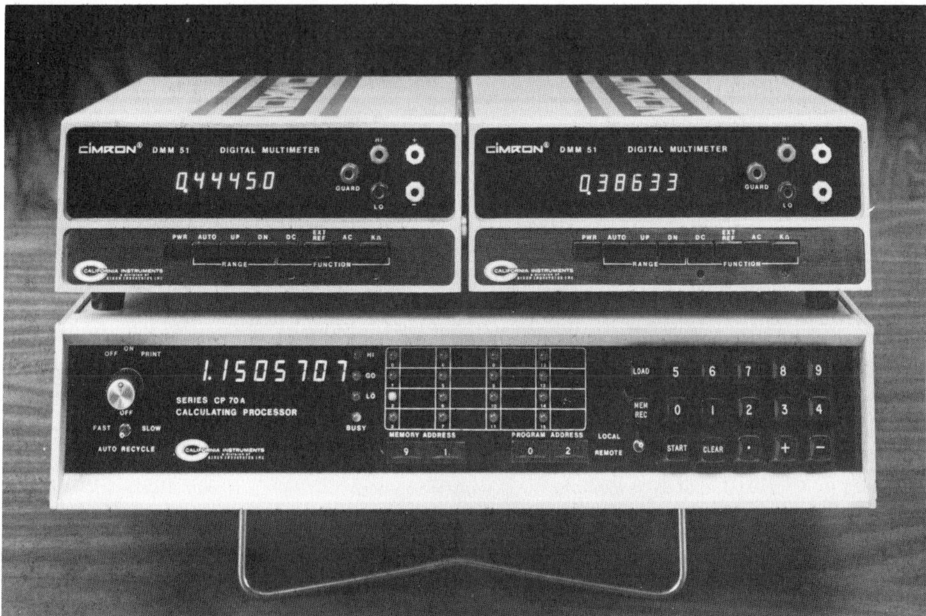

Figure 1.4 CP 70A-type programmed control unit by California Instruments

execute various measuring tasks and calculations sequentially.

The programming of the internally-programmed, computer-controlled instrumentation systems is determined by the programs stored in the memory which can be changed, thus allowing considerable flexibility. The fast-operating computer can also function with slow peripherals and other system units in a time-sharing operational mode.

Computer-controlled instrumentation systems are significantly more expensive than the previous category of systems. Figure 1.5 shows such an internally-programmed instrumentation system, the Type FAHD automatic analyser by Schlumberger. This particular instrumentation system can be used for testing the characteristics of VHF transmitters and is controlled by a DEC Type PDP 11/04 minicomputer.

1.3 Application areas

Automated measuring is a complex and expensive engineering task. The question arises, therefore, how to identify the advantages of automatic measuring systems over traditional, manually-controlled systems, and how to find the application areas where these advantages are most beneficial.

The advantages of automatic measurements are the following:

(1) *Greater speed*, due to the much faster operation of the control unit compared with a human operator, who can read the display of an analogue instrument at a rate corresponding to only about 10 bits/second. A digital instrument connected to a printer operates at around 1000 bits/second, while with a good quality magnetic tape recorder the figure is 1 Mbit/second. Due to the sampling principle governing the conversion of the analogue signals to digital format, the analogue signal can have only those frequencies that are lower than half the sampling frequency. As a consequence of this, the above-mentioned magnetic tape recorder can be used to record analogue signals of up to about 60 kHz only. If the signal contains harmonics higher than this frequency, then data errors during sampling are inevitable. In such cases a computer-controlled automatic measuring system equipped with a faster recording device is necessary, since their combined operating speed – according to the data transfer method – can exceed 1 Mbits/second. The great speed of the automatic measuring systems can also reduce the total time of the measurement. This is particularly significant when the measurement task is to be executed in multiple steps.

(2) *The measurement error is reduced.* The continuous increase in the demand for greater accuracy and growing complexity of the measurement tasks often means that they become impossible for the human operator. The results produced by

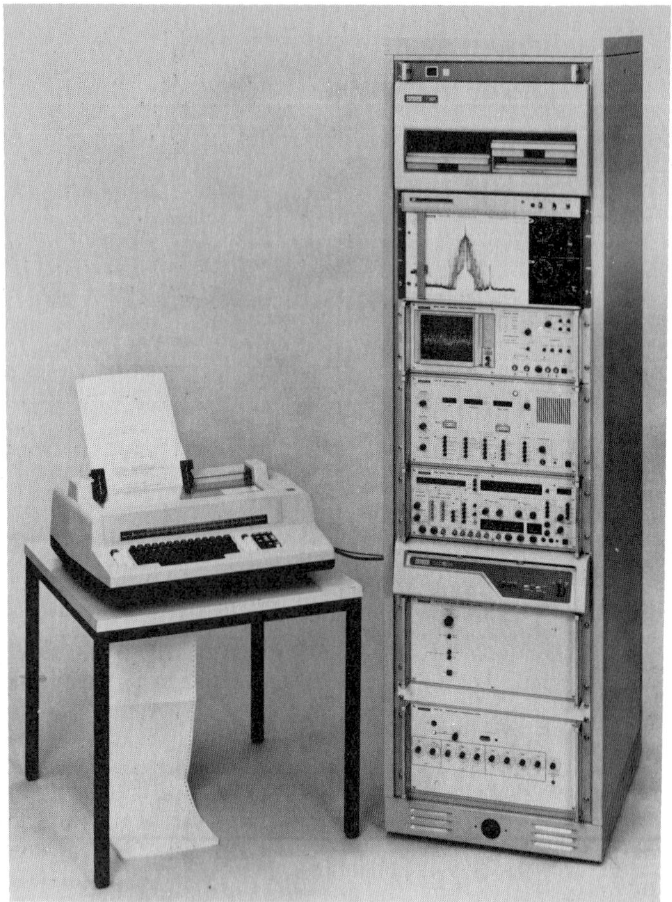

Figure 1.5 Type FAHD automatic analyser by Schlumberger

automatic measuring systems are free of subjective errors, due to the automatic control and the digital display of the results. A further advantage is that the system errors in an automatic measuring system can be determined, stored and considered in the evaluation of the results.

(3) *The results can be obtained directly in the appropriate engineering units and in an authentic certificated form.* This advantage is closely related to the earlier remarks concerning the reduction of measurement time. In some control units, mini-computers or calculators can handle the conversion processes to engineering units during the measurement and can calculate measuring probe positions, for example.

It is hardly necessary to stress the importance of result recording. With suitable peripherals connected to the control unit of the automatic system (e.g. serial printer), not only can measurement ledgers be made but the results necessary for further processing can be recorded on a data storage unit.

The picture would not be complete unless, as well as emphasizing the advantages, the disadvantages are also noted. During the design phase of an automatic measuring system the knowledge and the consideration of these can be of vital importance in ensuring the usability of the system.

The disadvantages of automatic measuring systems are the following:

(1) *The capital cost is considerably greater than that of a manually-operated system.* This greater cost has two components. First, consideration must be given to the cost of the control unit and programming, which does not exist for manual operations. Secondly, the cost of programmable, digital-signal-controlled instruments is greater than that of comparable manually-controlled types. These extra costs are less noticeable if the user constructs the automatic measuring system by utilizing standardized instrument interfaces. In this case, the controller can be an existing calculator or mini-computer and only the standard interface unit need

be purchased. Apart from this the capital allocated to the instruments may also be reduced, because many manufacturers now provide standard interface options on their own instruments.

(2) *There is a need for specially-trained personnel.* The designers and operators of automatic measuring systems must be familiar with relatively diverse fields of measuring technology and computing. Specialized training can often be provided by the manufacturers.

(3) *It is necessary to identify the measuring task accurately in detail before obtaining the automatic system.* With few exceptions, automatic measuring systems cannot as easily be adapted to changing measurement needs as can traditional, manually-operated systems. This is a serious disadvantage, as the extremely rapid technological development generates new demands on the measuring systems almost day-by-day.

(4) *Modifications of the units to be measured will be necessary for automatic measurements,* i.e. the special requirements of the automatic measurement must be considered in the design phase. The necessity of undertaking this at the outset is due to the difficulty of finding an appropriate substitute for the manual control when interfacing the unit under test. The characteristics of the changes are determined by the possibilities provided by the interface and by the nature of the measurement task. For instance, such tasks might include unusual test measurements of digital circuit panels, auxiliary outputs or gate circuits.

From the above it should be clear that the benefits of an automatic measuring system are realized when making measurements which:

- consist of a large number of steps or repeated operations;
- require simultaneous computing or decision-making operations;
- require great accuracy and reproduceability.

It is evident that measurement automation is not solely an engineering problem. In practice, the cost effectiveness of any proposed measuring system must be examined before making the decision to acquire or construct.

1.4 Interface systems

Various forms of automatic measuring system were produced as early as the mid-1950s. These were invariably custom-designed. One deciding factor against automation in those days was the considerably higher capital cost of automatic systems. This high cost was not solely the consequence of the more expensive instruments used in the automatic systems, as the price of the interface units necessary to connect the instruments and the hardware/software costs of the system also contributed significantly to the costs.

In the early development stages of digital technology the interfacing of automatic measuring systems took place by using parallel, binary-coded decimal (BCD) data outputs and remote-control functions and digital inputs. The data outputs (e.g. parallel output, BCD output, digital output, etc) were transferred in parallel bit form. These output signals were directly available on the outputs of the decimal counters of the instruments. For this reason the BCD output was an optional extra on nearly every digital output instrument made for laboratory use.

On such instruments the numerical value of the measurement was generated in 8-4-2-1 binary code. The remote-control function, however, varied widely in the individual instruments. Various manufacturers used different functions and, correspondingly, different remote-control signals. The only common feature was that generally 50-pin connectors were used on the back panel of the instruments for connecting the data and the remote control cables. The connection enabled the transfer of start, overflow and other similar signals in addition to the transfer of data and measurement limits. This approach facilitated the interfacing of the instrument and a printer that automatically registered the results.

Instruments specifically developed for automatic measuring systems made the remote control of some of the simpler functions possible, as well as the registration of measurement results. Selection of local/remote control, measurement limit change and operational mode are examples of these simpler functions. The signals initiating the control of the instruments were routed via a separate remote-control socket located in the back face of the instrument illustrated in Figure 1.6.

Only rudimentary systems could be made using the BCD interface, and the production of even these caused considerable financial and engineering problems for the user. In these costly and time-consuming, custom-designed systems the interfacing of the individual elements was unique for each system. Every instrument was connected to the control unit via its own data and control signal transfer cables. An additional problem arising from this was that any change in the measurement task, or any modification in the measuring requirements frequently necessitated redesigning a significant portion of the measuring system.

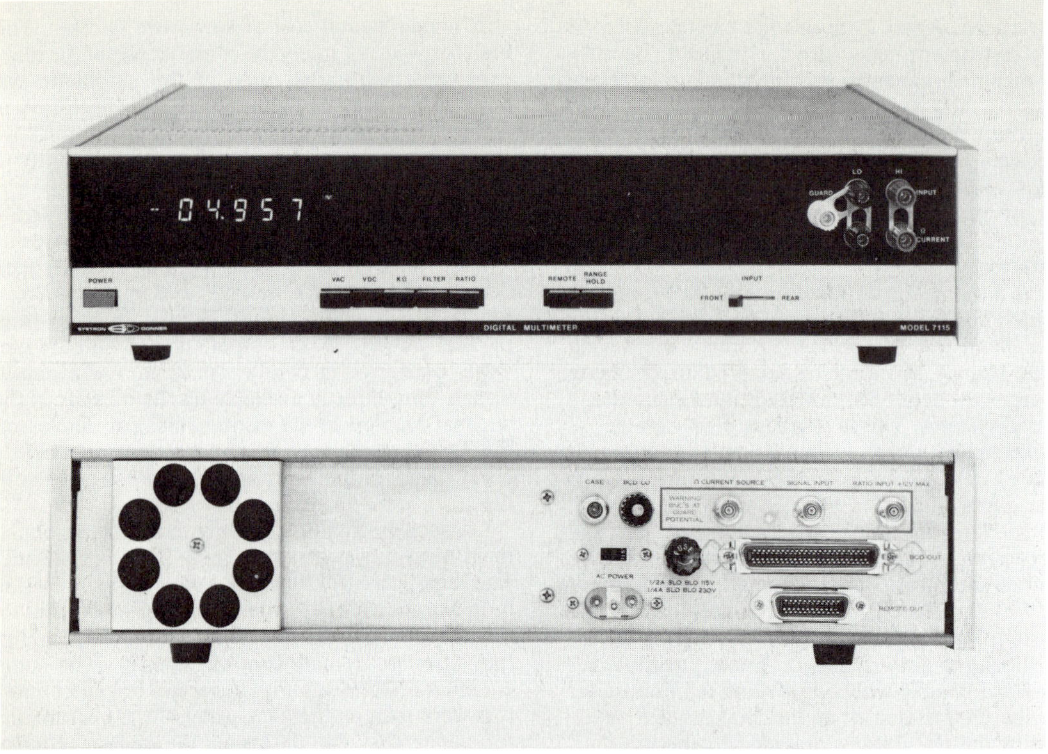

Figure 1.6 Two views of the Systron-Donner digital multimeter (Model 7115). The input/output connections providing data transfer are clearly visible on the back of the instrument

1.4.1 The RS-232-C interface

The performance of an automatic measuring system depends greatly on the operational speed of the computer used as the controller, or rather on the manner in which the fast-operating computer is interfaced to the individual devices.

The control units of measuring systems, the mini- or microprocessors, accept data to be processed in 8-, 12- or 16-bit parallel binary form. Hence a BCD/binary conversion is needed to interface the instruments. This in itself is not sufficient, as every computer has an interface requirement with precise timing specifications, usually different from other computers. For this reason an interfacing method that is compatible for all computers is required and the serial interface standardized for the Teletype peripherals of computers fulfils this requirement. It is described fully in the American EIA (Electronic Industries Association) Standard Application Note for RS-232-C standard and briefly in a later chapter. In this very simple, serial, asynchronous mode interface the standard ASCII (American Standard Code for Information Interchange) code system is used for the transfer of data and of some control commands. The interface is limited to a single data path and the data are transferred from the data source to a data receiver; therefore in a multi-device measuring system every individual instrument is connected to the computer via a separate interface.

1.4.2 Standard interfaces developed for measuring automation

Automatic measuring systems are still quite expensive, although there have been several important changes in this field, due to developments in technology that have influenced the trend of the performance:cost ratio of automatic systems considerably.

One significant change is that the difference in cost between manually-controlled and programmable instruments has decreased considerably. This is the direct consequence of the widespread use of microprocessor control. Many instruments now contain a microprocessor unit that can be used for control functions and for simple data processing.

Another advantageous change has taken place in the availability of high-level language (BASIC or HPL) programmable calculators (such as that

Figure 1.7 High-level language programmable desk calculator (Hewlett-Packard Type 9825)

illustrated in Figure 1.7) which are eminently suitable for the control of measuring systems. Although the performance and operational speed of such calculators falls short of that of computers, their programming is much simpler. The latter is a very important point, as the programming tasks present serious difficulties for the measuring technology expert.

The development of standardized interface systems was the most significant change in measuring automation. Towards the end of the 1960s, several measuring technology institutions began simultaneously the standardization of automatic measuring system interfaces. The objective of these interface recommendations and standards was to enable the interfacing of devices – supplied by various manufacturers for diverse applications – to each other and to the control unit, thus allowing easy expansion or modification of the resulting measuring system, as the measurement requirements changed. Several standard interfaces have emerged and these are described below.

1.4.3 The BSI interface

The British Standard Interface (BSI) system was amongst the first to be adopted as a standard (BS 4421:1969). This uses a one-way, serial byte, asynchronous data traffic operational system. The BSI interface is a 'point-to-point' connecting system, i.e. only two devices, the SOURCE and the ACCEPTOR, can be connected to the interface. The roles of the two devices cannot be interchanged, owing to the one-way data traffic; if a two-way data traffic is required, then two opposite direction BSI interfaces, complete with their interconnecting cables, must be provided between the two devices.

There are 18 signal lines in the BSI interface; of these eight are data lines, the remainder providing certain simple control functions. As a consequence of its structure described above, instrument addressing or interrupt signals are not present in the BSI specification. The BSI system has not become widespread in practical applications, primarily because the electronic parameters of the system were not compatible with modern integrated-circuit technology.

1.4.4 The SIAK and SIAL interfaces

The SIAK interface was developed within the COMECON countries and was initiated by East Germany around the same time as was the BSI interface. Similar to that of the BSI system, it was a simple structure. The organization and timing of data transfer are less restricted in this system, owing to an arbitrary word length and asynchronous operation. Even so, this system has not gained widespread application.

Another standard system is the SI 2.2/1974 interface, also developed in East Germany, which was accepted under the name of 'SIAL Standard Interface' in Section 8 of the Permanent Committee of Engineering of the COMECON.

This system was developed for data collecting and process control modular instruments. The modular structure is provided by the bus-like organization of the interface and modification or expansion of the operational system does not require hardware changes.

The basic structure of the SI 2.2 system, as shown in Figure 1.8(a), consists of a central unit and not more than 16 'blocks'. A common cable system interfaces the blocks to each other and to the central

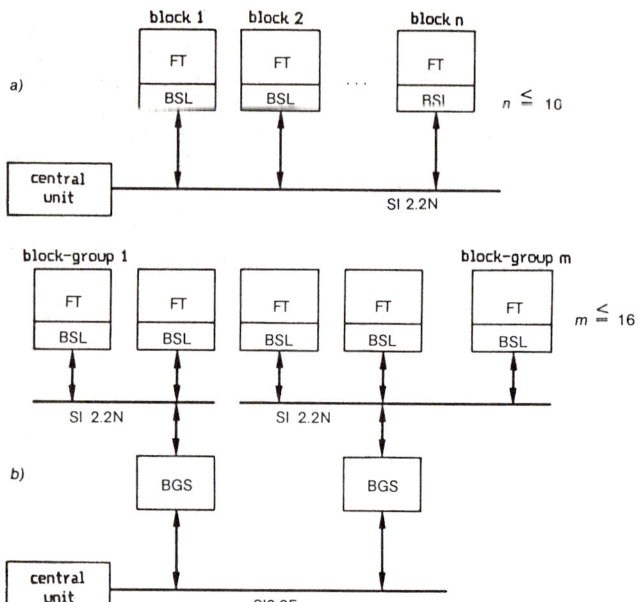

Figure 1.8 Structure of SIAL measuring systems. (a) Basic arrangement, (b) extended system

unit, permitting two-way information transfer. The process of information transfer is controlled by the central unit which, in most cases, executes the information processing as well.

The blocks consist of two parts, the operational section (FT) and the block control section (BSL). The latter interfaces the operational section to the common cable system. If more than 16 blocks are to be connected to the central unit, then the individual block groups are interfaced via a block group control unit (BGS) to the SI2.2F bus (Figure 1.8(b)). The information transfer between blocks is executed in parallel-bit mode; 16-bit-long information words can be transferred at a rate of 40000 words/second.

The SI2.2 system is used in computer-controlled processing systems and, in modified form, in the numerical control units of machine tools.

1.4.5 The CAMAC interface system

Another, more complex system is the CAMAC interface (Computer Automated Measurement And Control), developed during the 1960s by the multi-national ESONE Committee (European Standards Of Nuclear Electronics), within the EURATOM organizations for nuclear measuring systems. This culminated in a first specification document EUR4100, published in 1969, with others following. Collaboration was also maintained with the AEC NIM Committee in the USA.

The system has since been accepted by the American IEEE and the International Electrotechnical Commission (IEC). The former uses the IEEE Standards 583, 595, 596 and 683 together with other relevant documents concerning software and protocol; the latter uses publications IEC 582, 516, 552, 640 and 677 together with corresponding software publications.

The basic features of CAMAC are:

(a) it is a modular system, with functional plug-in units (modules) that mount in a standard crate;
(b) it is designed to exploit the high packing density possible with the solid state devices;
(c) the plug-in units connect to a common internal bus (Dataway) that is part of the crate and carries data, control signals and power;
(d) the system can connect to an on-line computer, although the use of a computer is optional;
(e) assemblies of crates may be interconnected by means of parallel or serial highways.

The system has the following data and address capabilities:

(a)	read lines (bus lines)	24
(b)	write lines (bus lines)	24
(c)	station (module) address (dedicated lines)	24
(d)	station (module) demand (dedicated lines)	24
(e)	subaddresses (per station address) (binary coded)	16

The Dataway has a minimum cycle time of 1μ second, and is a parallel highway carrying the 24 + 24 read and write lines from the modules to and from a controller, so that words up to 24 bits long can be handled in a single data transfer.

The plug-in CAMAC unit (module) is a functional unit that converts the external functions to conform to the Dataway requirements. Each module therefore terminates in a double-sided printed circuit plug of 86 contacts connecting to a corresponding Dataway socket. A module may be any multiple of the single-unit width of 17 mm, and the standard defines only dimensions, electrical power supply requirements, signal standards and data transfer protocol of the interface to the Dataway. The standard leaves complete freedom regarding the inputs to modules and the choice of internal functions and the internal circuit technology of each module is left to the designer.

The CAMAC crate serves as the common housing for the plug-in units and its dimensions, power requirements and Dataway are also specified in the standard. It provides accommodation for 25 module widths and the right-hand two locations of each crate are allocated to the crate controller (or interface to an external computer).

Due to these features, a wide variety of CAMAC modules with differing input features, each compatible with the system, are available from a considerable number of manufacturers.

CAMAC systems vary from simple single-crate systems – through relatively compact multi-crate systems – to complex widely-distributed multi-crate systems. These are specified by appropriate standards and have crate controllers of appropriate design. They are illustrated in Figure 1.9.

In Figure 1.9(a) an arrangement is shown, where the crate controller interfaces the crate directly to

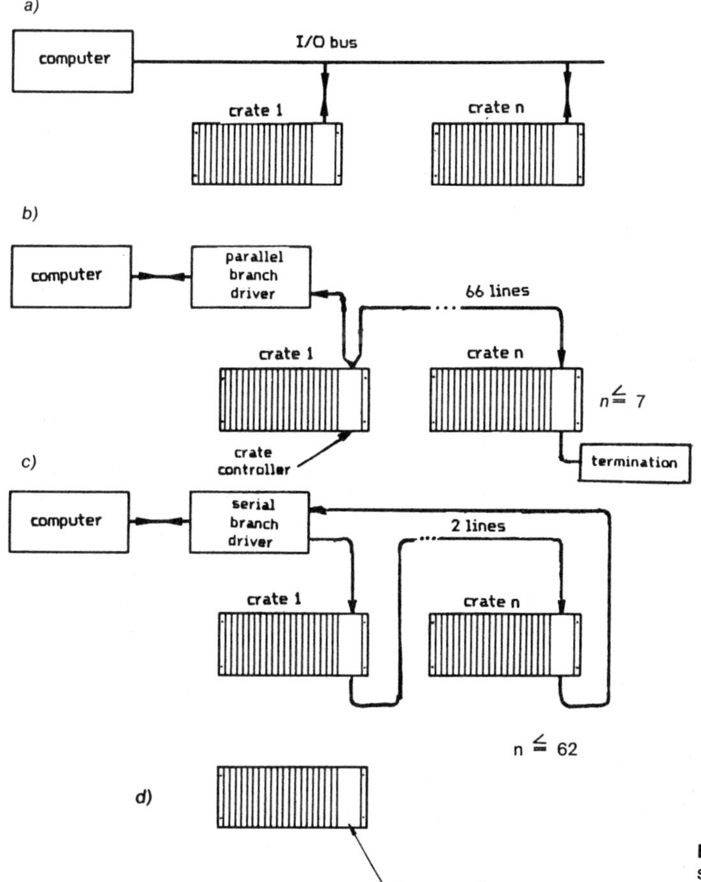

Figure 1.9 CAMAC variations. (a) Parallel system, (b) branch highway arrangement, (c) serial highway arrangement, (d) individual CAMAC crate

the parallel I/O bus of the computer. This arrangement permits the highest possible data transfer rate. but is suitable only for relatively short lengths of the I/O bus.

Figure 1.9(b) shows the branch highway arrangement that allows up to seven crates to be interfaced. The use of a branch driver permits the length of the highway to be extended (10–100 m) by the use of balanced transmission modes.

Where individual crates and the computer are several kilometers apart, or more than seven crates are necessary, the serial highway arrangement (shown in Figure 1.9(c)) is the appropriate one. With this method, up to 62 crates can be connected by either two balanced pair cables, operating in bit-serial mode, or by byte-serial mode, using nine balanced pairs. The distance can be extended in certain circumstances.

Micro- or minicomputers can also be utilized in CAMAC systems, either in individual modules or as crate controllers. An individual CAMAC crate is shown in Figure 1.9(d), where a microcomputer acts as the crate controller.

The main advantage of the CAMAC system is that the user is exempted from all design and interfacing problems in constructing the system, as the appropriately selected crates and modules, even if obtained from various manufacturers, can be simply connected together.

There are now more than 70 companies manufacturing some 600 CAMAC units and these are used in a wide variety of industrial process control applications, such as electricity generation and steel and aluminium manufacturing, as well as in research and development areas of the nuclear industry.

1.4.6 The IEC recommendation

The BSI, the SIAK and the CAMAC systems have not quite fulfilled expectations of providing a universally accepted solution to the problems of measuring automation standardization. One of the main reasons for this is that such systems are capable of meeting sophisticated requirements and therefore do not readily meet the requirements of the majority of users that a standard interface must be low in cost and offer simple operation.

The long time often required for developing and standardizing interface systems presents another problem. Owing to rapid technological development circumstances have been changing all the time and some standards were already out of date in many respects, when finally introduced.

It became apparent from the mid 1970s that the international standard for automatic measuring systems must be an interface system that is universally accepted by instrument manufacturers and users as well. This seemingly impossible task was finally solved by the International Electrotechnical Commission (IEC).

In the IEC Technical Committee 66 (which deals with programmable instruments), it was suggested by West German instrument manufacturers that a general purpose instrument interfacing system should be developed. When the suggestion was made, a basic condition was laid down that the new system should allow instruments generally available to be used with minimal modification.

The recommendation was well received by the large instrument-manufacturing firms and the first phase of work began in recording the universal requirements regarding standard interface systems. Exceptional foresight and painstaking deliberation were needed for this work. It was evident that the requirements concerning the new interface system could not possibly provide a solution for every existing measuring technology problem. The only aim could be that the new universal system should be applicable to the majority of measurement tasks in general practice.

The following basic limits were finally identified regarding the general purpose interface development:

- the maximum number of units in the system should be 15;
- the maximum rate of data traffic should be 1 Mbyte/second;
- the maximum distance between the units of the system should be 20 m;
- the length of the commands should be between 10 and 20 bytes.

A proposal for a system satisfying the above conditions was produced in 1972 by one of the largest instrument manufacturers, the Hewlett-Packard Company, and published in their own technical publication, the *Hewlett-Packard Journal*, in October, 1972.

This proposal, with slight modifications, was submitted to the Technical Committee 66 of the International Electrotechnical Commission in the September Conference held in Bucharest in 1974 and was accepted as an international recommendation.

The title of the recommendation is: 'Interface system for programmable measuring apparatus, byte-serial, bit-parallel'. (See reference in the Bibliography, Chapter 1.)

The symbol of the recommendation is:

The 120-page recommendation describes the electrical, mechanical and functional characteristics of the interface system. It states the requirements regarding the instruments to be used in the system and gives guidelines for designers to achieve measuring systems that are optimal in every respect.

The international recommendation was accepted by the standards organizations of several countries. Among others the American IEEE accepted it as Std 488-1975 and ANSI MC 1.1-1975.

The IEC recommendation has also been accepted by the COMECON countries. The corresponding COMECON standard bears the reference of IMR-2SI.

The IEC interface system was universally accepted by the large instrument manufacturers. As well as the originator Hewlett-Packard, other American firms like Fluke, Dana, Tektronix, Wavetek, Systron-Donner, Keithley, and European firms such as Philips, Siemens, Solartron-Schlumberger, Rohde & Schwarz, offer the IEC interface either as standard or as an option in their instruments (Figure 1.10). The instruction manuals of these IEC-interface-compatible instruments provide detailed instructions for programming and outline the important factors regarding their uses in measuring systems.

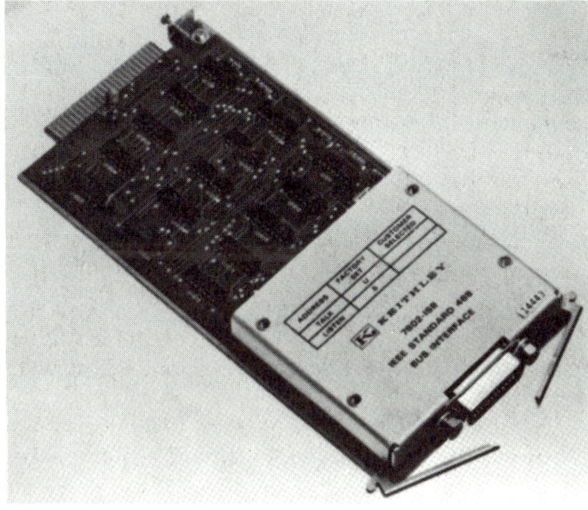

Figure 1.10 IEC interface unit by Keithley

Various manufacturers use different names for the interface system. HP-IB (Hewlett-Packard Interface Bus) is used by Hewlett-Packard, while GPIB (General Purpose Interface Bus), IEC 625, IEC Bus, IEEE-488, ASCII Bus and Plus Bus are also common names. (In this book we use the term 'IEC interface'. All details referring to this also relate to the interface systems marketed with the names listed above.) The difference in the name does not mean a change in practical implementation of the interface systems. Should a manufacturer deviate from the IEC recommendation, the loss of the greatest advantage of the standard interface will be inevitable, as its instruments could not be incorporated in systems with other manufacturers' IEC-compatible instruments.

Even the semiconductor manufacturers have taken notice of the widespread acceptance of the IEC interface and have been producing more and more special integrated circuits that facilitate the work of instrument designers.

The MC68 488, 40-pin, LSI integrated circuit made by the Motorola semiconductor manufacturer is suitable for implementation of the entire talker and listener functions of the IEC interface and, using some auxiliary circuits, even the control function can be implemented. Philips produce a similar integrated circuit under the reference HEF4738V, and there are also IEC chips available from Intel, Fairchild, Texas Instruments and NEC.

The universal international acceptance of the IEC interface obliged even those computer manufacturers who were not striving for standardization of any kind to make use of this interface system. Digital Equipment, Computer Automation, Tektronix, Intel, and of course Hewlett-Packard, are the main manufacturers of professional mini- and micro-computers and programmable calculators that can be used as measuring system controllers and that provide the IEC interface as a standard option. The interface can also be found in the most popular personal and home computers such as Commodore, Apple and others.

1.4.7 Further development of the IEC recommendation

Following the adoption of the IEC Publication 625-1, work continued in developing further specifications with a view to defining operational aspects of the interface system so as to offer a complete operating system.

The standard 625-1 specifies only the interface dependent messages; manufacturers of compatible instruments can therefore choose control and data transfer codes arbitrarily. Hence different, often very varied, commands are needed in programming the control units of systems constructed from instruments of different makes. Whilst it is not possible to define a single code set and message format which will satisfy a wide range of products using 625-1, it is feasible to define a limited set of guidelines regarding a number of different codes and formats of general applicability.

This has been done and published as IEC Standard, Publication 625-2 (IEEE 728), which states as its objects:

- to enable the highest possible degree of compatibility among different manufacturers' products;
- to enable the interconnection of apparatus;
- to generate, process and interpret a variety of different message types;
- to define codes and formats that will minimize the generation of application software and system configuration costs;
- to define a limited number of preferred message codes and formats in a relatively device-independent manner;
- etc.

Use is made of IEC 625-2 in subsequent chapters in the examples given of system configuration.

2 The IEC interface system

The IEC 625 Interface system (or IEEE 488) provides a standardized interface for programmable instruments. The following abbreviated description cannot replace the original IEC documents 625-1 and 625-2. It uses byte-serial, bit-parallel means to transfer data among groups of instruments and systems and is designed as an inter-device interface for devices in fairly close proximity to each other and able to communicate over a continuous party-line bus system.

It permits users of both simple and sophisticated, interconnected systems to configure easily devices from different manufacturers into one system.

2.1 Characteristics

Number of units connected: a maximum of 15 instruments.

Connection mode: a bus system, with passive connective cables; the interface functions are realized internally in the individual devices.

Connecting cables: are common screened cables, consisting of 16 signal lines and eight earth lines, their combined length not exceeding 20 m, or 2 m per device.

Plug connector: has 25 cylindrical contact pins and is trapezoid-shaped. Two preferred types exist according the European or the American standards and these are discussed later.

Data transfer: is byte-serial, bit-parallel, asynchronous.

Data rate: the maximum speed for the restricted cable length is 1 Mbyte/second; effective typical maximum data rate is 200–500 kbytes/second.

Typical message length: 10–20 characters.

Signal level: TTL.

Addressability: 31 talker and 31 listener addresses are possible for primary addressing (using 1-byte addresses).

2.2 The structure of the interface system

The interface described in the IEC 625 recommendation is a general purpose interface connecting the units of automatic measuring systems and providing a channel for the data traffic between them. The various system units (instruments, calculators, printers, etc), referred to as 'devices' in the following, are connected by the interface via a bus system.

The information transfer between the devices takes place in byte-serial, bit-parallel, asynchronous mode, in the form of so-called messages.

There are two kinds of message:

(a) *interface messages* that are responsible for the *system* operation and
(b) *device dependent messages* that are responsible for the operation of the *individual devices.*

2.2.1 Classification of the devices

The individual devices can have three different functions, with respect to the interface system.

- *Talker*. These devices can be selectively addressed via an interface message and in their addressed state they can generate and transmit device dependent messages to the bus. In a system only one device can be in the active talker state at any one time and the message on the bus can come only from that device.
- *Listener*. These devices can be selectively addressed via an interface message and in their addressed (active) state can receive device-dependent messages from the bus. In a system several devices can be in the listener state simultaneously.
- *Controller*. These devices are capable of selectively addressing the other devices in the system; in other words they can allocate the talker and listener roles and can transmit other interface messages to all, or only some addressed devices of the system. Controllers can function in a controller mode only, or also talk and listen. Those capable of control only do not transmit or receive device-dependent messages.

The roles of the devices in the operation of the interface system can be permanent or variable. Every device has at least one of the roles mentioned above; however, very often a single device can function in a different role at different times. For simplification, 'talker' will mean an active talker device in the following.

Every talker and listener device has an individual identification code called an 'address'. This address differentiates the device from the other units of the system. One device can have either a listener address or a talker address, or both. Within one system several devices can have identical listener addresses, provided that they always receive the same data. On the other hand, a system cannot contain two talker devices having identical addresses.

The number of devices connected in the IEC interface system cannot exceed 15. In a minimal configuration a single talker and a single listener device can be connected without a controller. An example for such a measuring system is a simple, semi-automatic data collecting system consisting of a digital voltmeter and a printer.

2.2.2 Bus structure

In the IEC interface system the individual devices are connected via a bus consisting of 16 signal lines. The entire data traffic of the interface system takes place in coded form on these 16 signal lines. In this bus system interface the main role of the connecting lines consists of making the parallel connections between the corresponding points of the identical type connectors of the individual devices.

The advantage of this interfacing is that expansion or configuration changes of the measuring systems are simple, which is of primary importance in a general purpose interface system.

The disadvantage of the parallel system is that quite complex addressing and control functions are required for data transfer between the individual devices. As a further consequence of the bus system, all devices must each have an interface unit. These convert the device-dependent messages (the instrument programming data, for example) arriving from the bus, into a format suitable for operating the device in question. The conversion of the internal signals of the devices (e.g. the measurement data) into a coded format appropriate for the bus system is also the task of these units.

The bus structure of the IEC interface system is shown in Figure 2.1. According to their roles in the operation of the interface system, the 16 signal lines are grouped as:

- data bus (eight signal lines);
- data transfer control bus (three signal lines), and
- interface administration bus (general control) (five signal lines).

The eight lines of the data bus transfer the interface messages in byte-serial, asynchronous operation mode. The message can be an address, programming data, measurement data or various commands and can originate from any device of the system that operates as a talker or a controller and is capable of transmitting the message.

The symbols designated for the data bus lines are:

- data input/output, DIO1 . . . DIO8.

The function of the data transfer control bus, consisting of three signal lines, is the control of the transfer of the data bytes on the DI0 signal lines, from an addressed talker, or from the controller to all listeners. The units of automatic measuring systems operate at various data transfer rates. Therefore, control will be necessary to effect the transfer of several-byte-long messages between the talker and the listeners. This control ensures that the talker will only 'talk' when it is being 'listened to' by the listeners, and that the effective operational

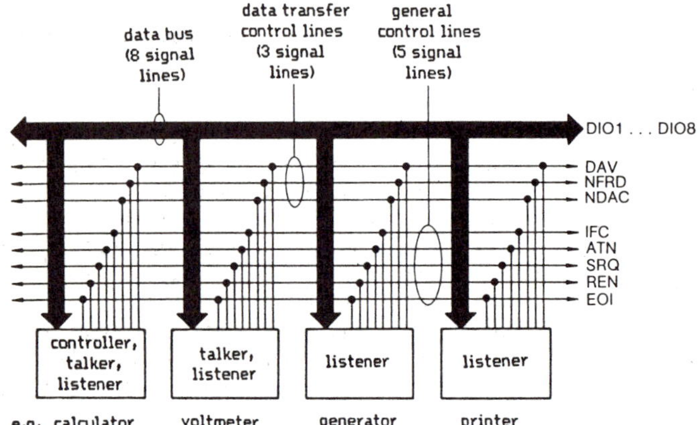

Figure 2.1 Structure of the IEC interface

speed of the talker will not exceed that of the slowest listener.

In the IEC interface system this control is the function of the asynchronous handshake cycle, taking place on the three signal lines of the data transfer control bus.

The symbols of these individual signal lines are:

- *Data Valid* (DAV): On this signal line level L indicates whether the information on the data bus is valid and acceptable;
- *Not Ready for Data* (NRFD): On this signal line the listener devices indicate (using level H) when they are ready to accept data;
- *No Data Accepted* (NDAC): On this signal line the listener devices indicate (using level H) the acceptability of data.

The data transfer control signals are produced by the interface units of the talker and the listener devices taking part in the data traffic, and not by the control unit.

All five signal lines of the interface administration bus each have an individual task in the control of the organized flow of information across the interface:

- *Interface Clear* (IFC): The clear signal line of the interface system is used by the controller to reset all devices of the system to the start state. This ensures identical conditions for the start of all interface operations. In systems containing several controllers this message can be given only by the highest priority controller, the so-called 'system controller';
- *Attention* (ATN): On the 'attention' signal line the controller specifies how the information on the data bus is to be interpreted and which device must respond. If there is level L on this signal line, then there is an interface message on the data bus – that is to say an address, a command or an address command, while in the case of level H there is a device-dependent message (e.g. program command);
- *Service Request* (SRQ): Devices can indicate the need for attention and ask the controller to interrupt the current operation by means of this signal line;
- *Remote Enable* (REN): On this signal line the system controller can, together with other messages, convert a unit to remote control operation, or select between two alternative sources of device programming data;
- *End Or Identify* (EOI): The last unit of a multiple-byte data group can be indicated by this signal line – in this case originating from the talker device. This line, together with the ATN, ensures the execution of parallel poll; in this case the message is generated by the controller.

2.3 Electrical and mechanical specifications

2.3.1 Electrical specifications

The use of TTL (Transistor Transistor Logic) technology is presumed in all standards related to the driver and receiver circuits of the IEC interface system. Apart from these, the circuits providing the interface functions can be built using other technologies, according to the preferences of the designer.

2.3.1.1 Signal levels

In the message code table the relationships of the given logic states and the measured electrical levels on the signal lines are the following:

Logic value (code)	*Electrical signal level*
0 (false)	≥ + 2 V (called the 'high state')
1 (true)	≤ + 0.8 V (called the 'low state').

The voltages above must be measured on the signal line relative to logic ground, at the connectors of the devices.

The circuit used for connecting a device to a particular signal line is shown in Figure 2.2. In the devices every signal line must be terminated with a resistive type load, even when the driver or receiver circuits are not connected to them. The purpose of

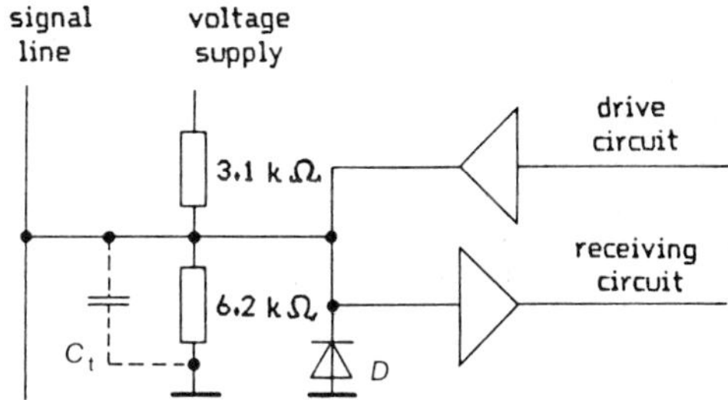

Figure 2.2 Circuit diagram of the IEC bus signal line connections

such terminations is that a pre-determined voltage value should be present on the signal line, even when all driver units connected to it are at level H. Apart from this, the termination ensures uniform device impedance and reduces susceptibility to noise. The maximum value of load capacitance (C_t), shown in the Figure is 100 pF per device.

Every device must contain a circuit limiting the negative voltage transients. The diode, (D), incorporated in the receiver circuit, as shown in the Figure, serves such a purpose.

2.3.1.2 *Loading*

The loading characteristics of the devices are determined by the resistive terminations, the voltage-clamping circuits and the driver/receiver circuits. Hence the standard load characteristics are not given for the individual components but for the entire device in the interface plane (Figure 2.3). For correct operation the DC load characteristic, determined by the voltage of the signal line and the input current, must fall within the unshaded area.

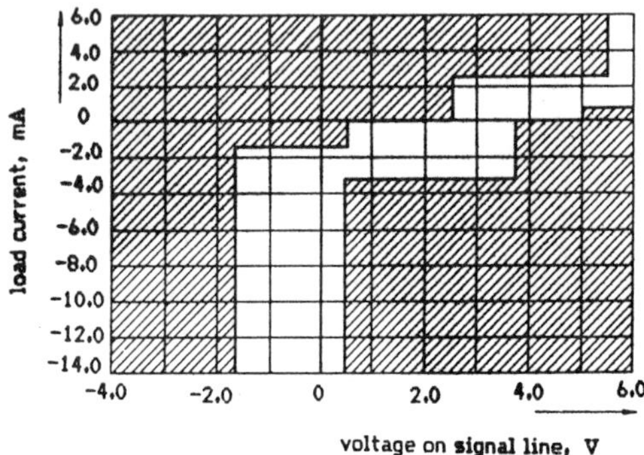

Figure 2.3 Load standards of the IEC interface

2.3.1.3 *Cable construction*

In the IEC interface system the total length of cable connecting the individual devices must not exceed 20 m. Cable lengths can now be extended by use of a proprietary bus extender. This converts the parallel data into serial bits suitable for transmission over twin-pair, coaxial or fibre optic cables or through a modem link. For such length cables, appropriate shielding against the reception or generation of electrical noise and interference becomes necessary and the IEC standards therefore give a detailed description of the cable requirements.

The connecting cable used in the interface system must at least contain an external shield and 24 conductors. Of these, 16 conductors serve as signal lines, the rest being used as logic ground returns. Each of the NRFD, DAV, NDAC, EOI, ATN and SRQ signal lines must be twisted with a ground lead or equivalently shielded to minimize crosstalk.

2.3.1.4 *Driver circuits*

The driver circuits of the signal cables are either open collector or three-state drivers. The use of the latter is justified in high-speed systems.

Only open collector drivers can be used for the driving of SRQ, NRFD and NDAC signal lines, because these signal lines connect the appropriate points of the individual devices with a wired-OR connection (see Section 2.4).

Open collector, or three-state drivers are equally suitable for driving the DAV, IFC, ATN, REN and EOI signal lines.

Only open collector drivers can be used for driving the DIO1 . . . DIO8 lines, when there is a need for parallel polling to be carried out. If there is no such need, then either type of driver can be used.

2.3.2 Mechanical specifications

The only area in the standard that has uncertainty concerns the connectors that are attached to the devices and to the cables.

Hewlett-Packard, the originators of the interface system, used the 24-pin Amphenol or Cinch Series 57 connector in their instruments and calculators from the very beginning. The same type of connector was accepted by the American IEEE and ANSI Standards Committee and is correspondingly used by American instrument manufacturers.

This robust, successful and practical connector, however, is not included in the range of the IEC-recommended connectors and therefore the IEC recommendation could not include it either. Instead, a similar connector, the 25-pin, trapezoid-shaped, cylindrical contact – included in the IEC 48B (Secretariat) 80A recommended list – was chosen. Figure 2.4 shows the schematic diagram and pin allocation of this connector.

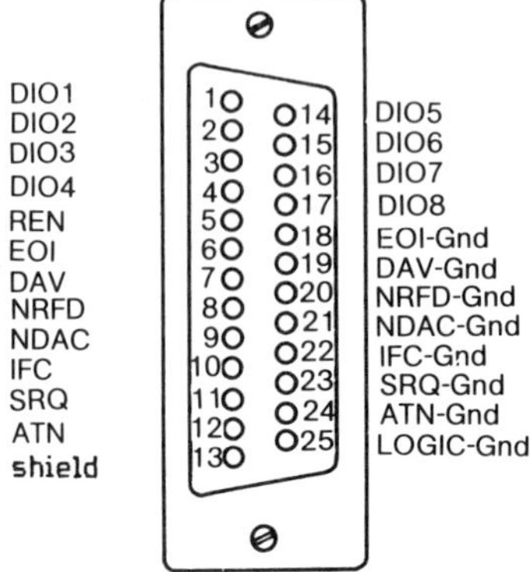

Figure 2.4 Wiring of the standard IEC interface connection

Both the IEC and IEEE specifications require a combined plug-and-socket-type connector to be fitted to both ends of the cables in back-to-back fashion. In this way individual cables can be plugged

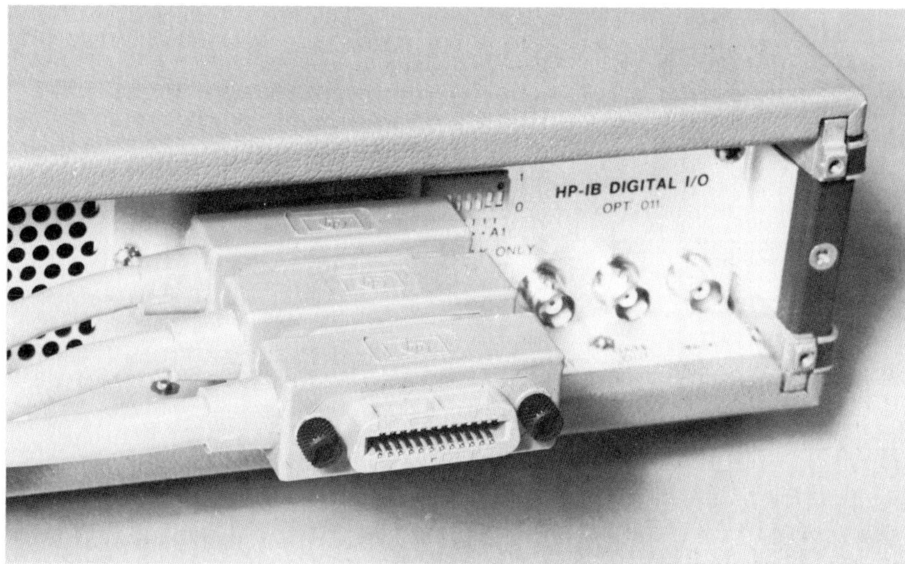

Figure 2.5 Multiple (piggyback) connection on the rear of a Hewlett-Packard instrument

into each other (piggyback connection), thus facilitating the connection of devices in either serial or star systems (Figure 2.5). The device must always be fitted with a plug type connector.

2.4 The handshake process

The maximum 1 Mbyte/second speed data traffic of the IEC interface is co-ordinated by the three-line data transfer bus, or in other words, by the handshake process taking place on them. If a talker sends data via the interface to the listener(s), then the rate of data transfer is determined by the operational speed of both the talker and the listener(s).

As the interface system provides a connection for differing measuring systems and, as the operational speed of these instruments may differ widely, the data transfer can only take place asynchronously. This allows even the 'slowest' listener to receive data

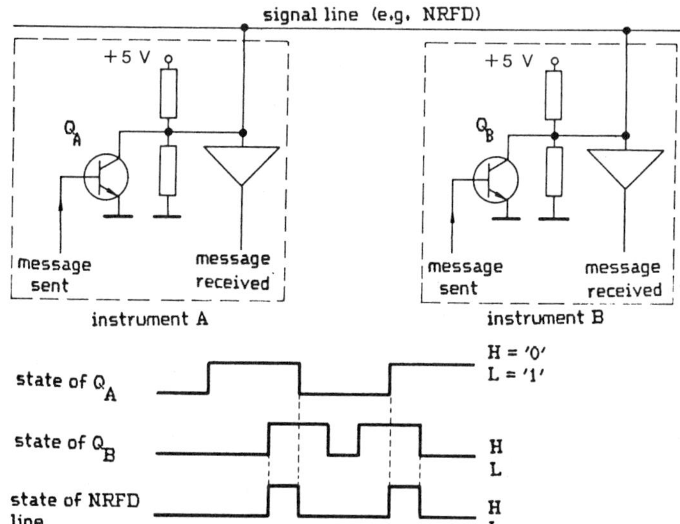

Figure 2.6 Wired-OR connection of the signal line driver circuits

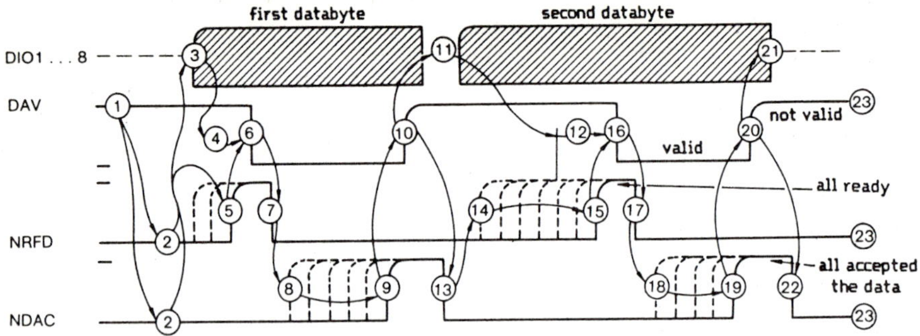

Figure 2.7 Timing of the handshake process

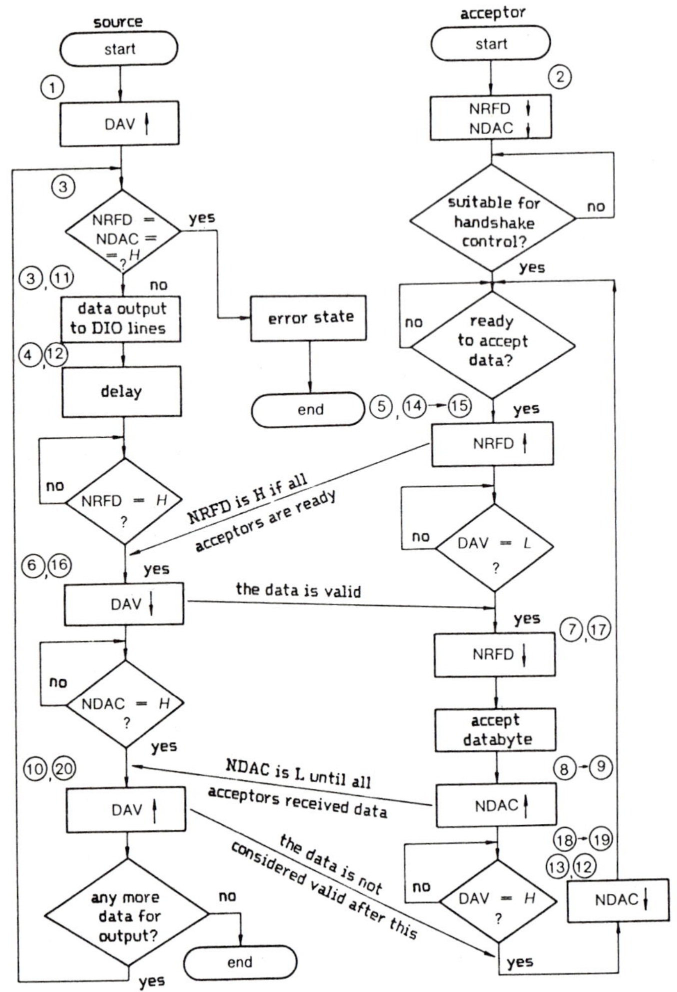

Figure 2.8 Flow diagram of the hand: process

from the 'fastest' talker, as the data transfer takes place at a speed determined by the 'slowest' device.

The operation of the three handshake lines is based upon the use of the wired-OR function, well known in digital technology. Figure 2.6 shows the connection of one of the signal lines (NRFD) of the bus, controlling the data transfer to a device. The open collector driver circuits on the device outputs operate as conducting connectors to earth; therefore the NRFD signal line will be at level L if either one or both of the drivers (Q_A and Q_B) is switched on. As a consequence of the negative logic convention of the IEC bus system, the NRFD is true (at level L) when either of the device outputs connected to the line is at level L.

In the IEC interface system the data transfer controlling bus creates a wired-OR logic function between the NRFD and the NDAC outputs of the individual devices. For this reason level H is present on the signal lines only when all active listener devices (including the slowest) indicate with output level H that they are ready to accept the new databyte (NRFD), or have already done so (NDAC).

The DAV signal line of the data transfer control bus is controlled by the source (talker), while the NRFD and the NDAC signal lines are controlled by the receptor devices (listeners).

The timing functions of the handshake data transfer cycle are shown in Figure 2.7, while Figure 2.8 illustrates the flow diagram of the operation. The circled numbers on both Figures denote the same events:

1 The source resets DAV to the original H state, thus indicating that the data is not valid.
2 The receivers reset the NRFD and NDAC lines to the original L state.
3 The source checks whether the NRFD and NDAC are at the L level and then transmits the databyte to the DIO lines.
4 The source sets up a delay to allow the data on the DIO lines to stabilize.
5 All receivers have signalled that they are ready to accept the databyte (the NRFD line will be at level H).
6 The source having sensed that line NRFD is at level H, sets the level of the DAV line to L, thus indicating that the data have stabilized and are valid.
7 The first receiver sets the NRFD to L state, to indicate that it is not ready now, then receives the data – the rest of the receivers act similarly at their own operational speed.
8 The first receiver sets the NDAC to H state to indicate that it has accepted the data (at this time, owing to the wired-OR connection, level L still prevails on the NDAC).
9 The last receiver has accepted the data as well, resulting in level H on the NDAC signal line.
10 Sensing the H state of the NDAC, the source cancels the data valid signal (the DAV will be at H level).
11 The source switches a new databyte onto the DIO lines.
12 The source sets up a delay to allow stabilization of data on the DIO lines.
13 Sensing the H level of the DAV (10), the receivers reset line NDAC to level L in preparation for the next cycle.
14 The first receiver signals on the NRFD line with H state that it is ready to accept the next databyte (the NRFD remains at level L).
15 The NRFD signal line will be at level H simultaneously with the last receiver becoming ready.
16 The source indicates validity of the data with level L on DAV.
17 The first receiver indicates the start of data transfer with level L of NRFD.
18 The first receiver indicates with level H on the NDAC line that it has accepted the data.
19 The last receiver accepts the data; thus the NDAC line will be at level H as well.
20 The source indicates data validity with level H on DAV.
21 The source then 'removes' the databyte from the DIO lines.
22 The receivers prepare to accept the next databyte setting NDAC to level L.
23 The handshake process returns to the original state (1).

It is apparent from the timing diagram that both the NRFD and the NDAC signals are complex, consisting of composite signals from two or more receivers. This is a consequence of the different operational speeds of the listeners, which do not respond simultaneously, because of the difference in the data transfer path lengths and different response times.

However, the data transfer between two fast devices is not slowed down by devices that are physically connected to the bus but are not taking part in the data transfer cycle (i.e. not addressed). There is constant H level on their NRFD and NDAC lines and they therefore do not affect the data transfer.

It should be noted that Hewlett-Packard hold patents in many countries on the three-line handshake technique. On payment of a nominal sum, companies are entitled to use the principle without disclosure of the applications.

2.5 Functional structure of the interface system

2.5.1 Functional elements

In Figure 2.9 the functional elements of an IEC compatible device are shown. The functions can be divided into three groups:

- device functions,
- interface functions, and
- message coding.

Essentially, the device function is the application that the device or instrument was designed for; e.g. voltage measurement in the case of a voltmeter, recording of measurement results in the case of a printer, etc. The standards of the IEC interface system do not deal with the disposition, contents and structure of the device functions. This ensures the general purpose, universal applicability of the interface. The instrument designer can freely determine the capability of the device and can choose the way to realize these features. This functional part is denoted by B in the Figure.

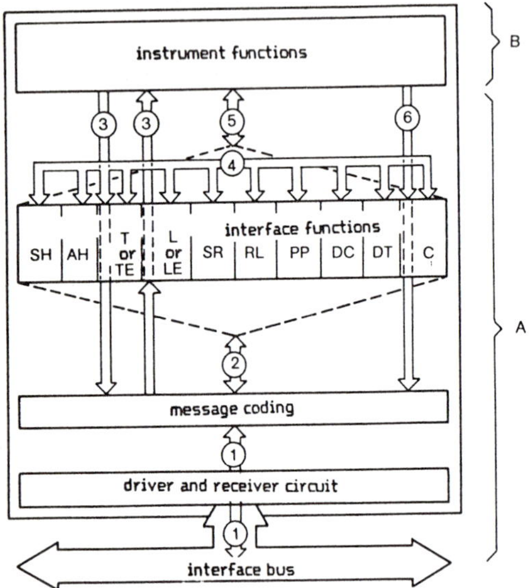

Figure 2.9 Diagram of the IEC-interface-compatible instrument functions

The interface functions are those elements of the device's operation that allow the information exchange with the other devices in the system; in other words the reception, processing and sending of messages.

The IEC system contains the following ten interface functions:

- source handshake (SH),
- acceptor handshake (AH),
- talker, or extended talker (T or TE),
- listener, or extended listener (L or LE),
- service request (SR)
- remote/local control (RL),
- parallel poll (PP),
- device clear (DC),
- device trigger (DT) and
- controller (C).

This so-called 'interface function set' is precisely defined by the IEC interface standard. The instrument designers cannot define new interface capacities and therefore the interface set cannot be expanded. On the other hand, the necessary interface functions can be freely selected during the instrument design stage as the actual use of the device will determine which of the interface functions are necessary.

The circuits of the interface functions can be devised freely, provided that the conditions relating to every individual status of the interface functions are complied with. The interface functions, their states and the transitions between them are discussed in detail in the next Section.

The message coding is a two-way function, partly in connection with the information arriving to and from the interface functions. The interface functions and all information transfer in their vicinity are carried out by transmitted or received messages. Two stages of the messages can be distinguished and any message can have a true or a false status at an arbitrary moment. The use of this system is justified because in the case of multiline messages, increased complexity and lack of clarity would occur if the status of message transmissions over parallel signal lines were given by the H and L levels. The use of true and false values allows simple binary evaluation even for multiline messages.

There are two basic kinds of message group, the local and the remote messages. Local messages are between the device and the interface functions. The incoming messages to the interface are defined and a new local message cannot be introduced into the interface functions. On the other hand, the outgoing messages from the interface can be chosen by the designer, i.e. a local message originating from any state of an arbitrary interface function can be incorporated into one or more device functions. The messages between the interface functions of different devices are called 'remote messages'. As was mentioned in the introduction to this Chapter, there are two kinds of remote message: interface messages and device-dependent messages.

In the individual devices the ATN message avoids ambiguity in the evaluation of messages. If the ATN message is false (there is level H on the ATN signal line), then the devices interpret the message as a device-dependent one. In this case the message can be received by the active listener devices only. The device-dependent messages are those between the device functions and the message coding. For example, programming data or measurement data are such device-dependent messages (the messages denoted by '3' in Figure 2.9). If the ATN message is true (there is level L on the ATN signal line), then the devices interpret the message as an interface message. Interface messages can be received and processed by all devices in the system. Among others, the talker and the listener addresses and general commands are such messages. In every case, the purpose of the interface messages is to bring about a state change within the interface function (messages denoted by '2' in Figure 2.9).

The messages can be classified according to the number of signal lines used for their transfer. Messages sent out on one signal line are called 'uniline messages'. The IEC recommendation permits the simultaneous transmission of two or more uniline messages; naturally on different signal lines. The transfer of multiline messages occurs on a signal line group. Only one multiline message can be transferred at any one time.

2.5.2 Interface functions

In the IEC interface system ten interface functions serve as generators and processors of the interface messages. Basically these functions ensure the universal character of the interface system and determine its capability.

Figure 2.10 shows the interface functions, the message coding and the signal line system transferring the messages. As is apparent from the Figure, the task of the interface functions and the message coding is to connect the individual internal control system of the device to the precisely-defined, unified system of the IEC interface.

The interface functions can be classified in two groups. The five so-called 'basic functions': the

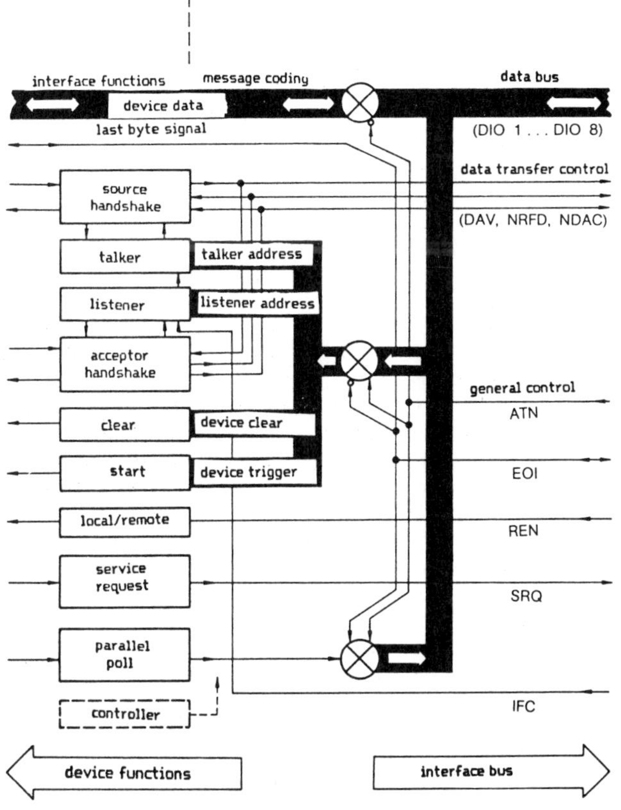

Figure 2.10 Interface functions of the IEC system and message coding

talker, the listener, the controller, the source handshake and the acceptor handshake serve the basic data traffic. The service request, the remote/local control, the parallel poll, the device clear and the device trigger (start) are auxiliary functions.

In the devices the basic functions are used more frequently. All devices have a talker or a listener function, or maybe both. Only a few devices have a controller function, although there are many instruments that have a limited control capability. All devices must have a source or an acceptor handshake function, because without these data traffic is not possible.

In the IEC interface system the description of the interface functions is carried out with the aid of state diagrams. This method ensures the unambiguous definition of the functions and facilitates the realization of the desired function with the use of an arbitrary circuit technology. For this reason the interface functions are entirely universal with respect to both their application and their realization.

2.5.2.1 Coding system

In the IEC recommendation all interface functions are defined by a so-called 'state system', consisting of several mutually-exclusive states. Every state is symbolized with a circle on the state diagrams defining the individual functions. The upper case four-letter symbol written in the circle identifies the state (Figure 2.11(a)). These state symbols are derived from the initial letters of the English name of the state; for example, as shown in the Figure, 'SIWS' stands for 'Source Idle Wait State'. The state symbols always end with an 'S', denoting the initial letter of 'state'.

The possible transfers between the interface

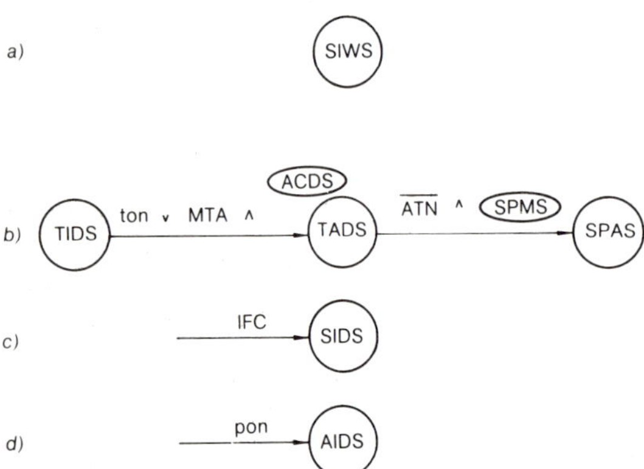

Figure 2.11 Interface function state symbols. (a) State symbols, (b) state relation symbols, (c) abbreviated symbol for interface clear, (d) abbreviated symbol for start mode

functions are represented by arrows. Every transfer is qualified by an expression, whose value is either true or false. The interface function changes its state if an expression, relating to the transitions starting from a given state, becomes true. The expressions characteristic to the transitions consist of local messages, remote messages and state relationships, or minimum time limits used in conjunction with the operators AND, NOT and OR (Figure 2.11(b)).

Local messages are represented with a lower case three-letter symbol which consists of the initial letters of the local message; for instance, 'pon' stands for 'power on'. The local messages of the interface are summarized in Table 2.1, which also shows the appropriate interface functions. The individual interface functions are denoted by their initial letters.

The remote messages arriving from the interface are denoted with an upper case three-letter symbol, using their initial letters; for example, 'MTA' stands for 'My Talk Address'. The remote messages of the IEC interface are summarized in Table 2.2.

Table 2.1 Local messages of the IEC interface

Abbreviation	*Message*	*Function*
gts	go to standby	C
isr	individual service request	PP
lon	listen only	L, LE
(lpe)	local poll enable	PP
ltn	listen	L, LE
lun	local unlisten	L, LE
nba	new byte available	SH
pon	power on	SH, AH, T, TE, L, LE, SR, RL, PP, C
rdy	ready	AH
rpp	request parallel poll	C
rsc	request system control	C
rsv	request service	SR
rtl	return to local	RL
sic	send interface clear	C
sre	send remote enable	C
tca	take control asynchronously	C
tcs	take control synchronously	AH, C
ton	talk only	T, TE

The so-called 'state relationships' between the individual interface functions are denoted with an upper case four-letter symbol enclosed in an ellipsoid shape. This symbol is the abbreviation of the name of the interfacing state, for example

Table 2.2 Remote messages of the IEC interface

Abbreviation	*Message*	*Function*
ATN	attention	SH, AH, T, TE, L, LE, PP, C
DAB	databyte	(via L, LE)
DAC	data accepted	SH
DAV	data valid	AH
DCL	device clear	DC
END	end	(via L, LE)
GET	group execute trigger	DT
GTL	go to local	RL
IDY	identify	PP, L, LE
IFC	interface clear	T, TE, L, LE, C
LLO	local lockout	RL
MLA	my listen address	L, LE, RL
(MLA)	my listen address	T
MSA	my secondary address	T, LE
(MSA)	my secondary address	T, TE
MTA	my talk address	T, TE
(MTA)	my talk address	L
OSA	other secondary address	TE
OTA	other talk address	T, TE
PCG	primary command group	TE, LE, PP
PPC	parallel poll configure	PP
(PPD)	parallel poll disable	PP
(PPE)	parallel poll enable	PP
PPRn	parallel poll response n	(via C)
PPU	parallel poll unconfigure	PP
REN	remote enable	RL
RFD	ready for data	SH
RQS	request service	(via L, LE)
(SDC)	selected device clear	DC
SPD	serial poll disable	T, TE
SPE	serial poll enable	T, TE
SRQ	service request	(via C)
STB	status byte	(via L, LE)
TCT	take control	C
(TCT)	take control	C
UNL	unlisten	L, LE

'SPMS' means 'Serial Poll Mode State'. The state relationship is true when the signalled state is active; otherwise it is false.

The AND operator is represented by the '∧' symbol in the expressions describing the transitions of the interface functions. The operator has the same effect as the Boolean AND function. Similarly the '∧' symbol represents the OR operator, which has an effect identical to that of the Boolean OR function.

In the expressions the NOT operator is used as well, represented by a horizontal line above the portion of the expression to be negated. The value of the negated expression is true only if the expression under the horizontal line is false.

It has been discussed already that the interface functions were developed to avoid timing problems in instrument design, even when different circuit solutions are chosen by the designers in the individual instruments. Hence, certain time limits must be adhered to in order to achieve optimal compatibility of the instruments. The values for these time limits are displayed in Table 2.3. They originate from the propagation time of the transition and from the delay time of the circuit characteristics of the devices. The values in the Table are valid for open collector driver circuits, and they are different for the faster-operating three-state drivers.

The time limits are shown on the state diagrams of the individual functions, and are represented by two kinds of symbol:

- The so-called 'lower time limits', denoted by T_n, meaning the minimum time a function must remain in a given state.
- The so-called 'upper time limits', denoted by t_n, meaning the maximum time allowed for a state transition.

Table 2.3 The values of the time limits appearing on the state diagrams of the interface functions

Symbol	*Relevant function*	*Description*	*Time limit*
T_1	SH	stabilization time of multiline messages	⩾2 μsecond
t_2	SH, AH, T, L, LE, TE	response to ATN	⩽200 nsecond
T_3	AH	acceptance time of an interface message	>0
t_4	T, TE, L, LE, C, RL	response to IFC or REN is a false message	<100 μsecond
t_5	PP	response to ATN EOI messages	⩽200 nsecond
T_6	C	parallel poll execution time	⩾2 μsecond
T_7	C	controller delay to enable talker to recognize ATN	⩾500 nsecond
T_8	C	length of IFC or REN false message	>100 μsecond
T_9	C	delay for the EOI message	⩾1.5 μsecond

In addition, there are a few special symbols on the state diagrams of the interface functions, as follows:

- If a certain part of an expression can be arbitrarily selected, and therefore does not have to be true to ensure that the whole expression is true, then this part is enclosed in square brackets, e.g. [MLA].
- If an expression brings about a transition to a particular state from every other state on the diagram, then an abbreviated symbol is used instead of representing every transition. This situation is represented by an arrow not having an original state (Figure 2.11(c)). This kind of symbol can be found generally at the start state of the functions.

There is an additional general abbreviation that is used in connection with the start states. For simplicity, the switched-off state of the devices is not shown on the state diagrams. Figure 2.11(d) shows the abbreviated symbol. The 'pon' message sets the first state defined at switch on.

2.5.2.2 *Definition of the IEC functions*

Unambiguous definition of the interface functions ensures the compatibility of IEC systems. Hence, the IEC recommendation describes these functions very precisely and unambiguously with the symbol system described briefly above.

The basis of the description is the state system consisting of one or more state diagrams per interface function. The state diagram is augmented by textual description of it and by tabulated information on the individual states and the messages generating those states. Finally, the IEC recommendation also contains, in tabulated form, the outgoing messages of the individual functions and the possible variants of the functions.

In the following, the interface functions will be described in a considerably simpler and shorter manner than in the IEC recommendation. This descriptive manner will be naturally more ambiguous and cannot replace the IEC recommendation in any way.

The state diagram will be described in detail only for the talker and listener functions; the reader can interpret the rest of the state diagrams relatively easily in an analogous manner. However, the names of the states of all functions will be given in tabulated form, as well as the names of the possible variants of the functions and, briefly, their characteristics. The main advantage of the IEC interface system is that it is not necessary to build every specified function into each device. Moreover, individual functions can be built in only partially, according to the requirements of the device's operation.

2.5.2.3 *Talker (T) and Extended Talker (TE) functions*

A device possessing the talker function can send device-dependent data to another device via the interface, but this capability exists only when the given function is addressed to talk.

There are two variants of the talker function, differing only in their addressing modes: one with and one without address extension. The normal function (T) uses a one-byte address, while the address of the extended talker interface function

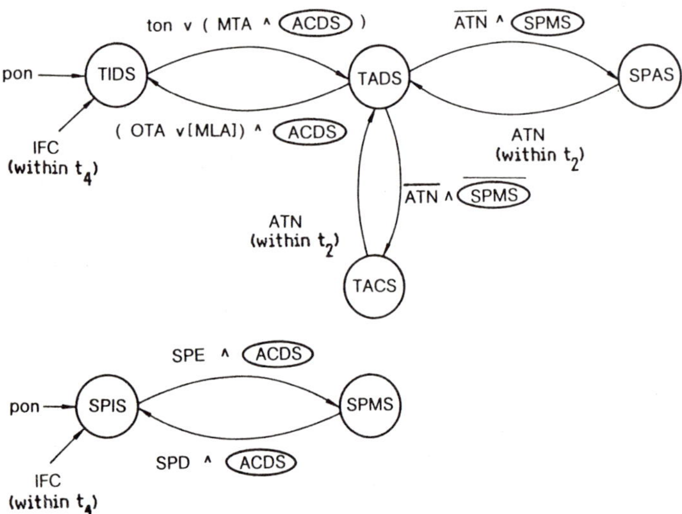

Figure 2.12 State diagram of the talker function

(TE) consist of two bytes. The capabilities of the two variants are identical in every other respect. On any device only one of the two talker functions can be realized.

The state diagram of the talker function is shown in figure 2.12. The talker function can be in two active states simultaneously, provided that the two states belong to separate sets. The corresponding states of the two state sets shown in the Figure are mutually exclusive and therefore only one state from each set can be active. This condition applies equally to the state diagrams of the other interface functions.

The names of the various states of the talker function are summarized in Table 2.4. 'TIDS' is the 'Talker Function Start State', when the power supply of the device is switched on, or when an Interface Clear message (IFC) is received. According to definition, the talker function must return to the TIDS state from any other active state when an IFC message is received. In the TIDS state the talker function cannot transmit data or state bytes.

Table 2.4 The states of the talker function (T)

Abbreviation	*State*
SPAS	serial poll active
SPIS	serial poll idle
SPMS	serial poll mode
TACS	talker active
TADS	talker addressed
TIDS	talker idle

If the IFC message is false and the switch executing the 'talker only' (ton) message is in the 'on' position, or the device receives its own talk address (MTA) simultaneously with the ACDS state of the receptor handshake function, then the talker function must go to the TADS state. In this state the talker function had already received its own address and accordingly prepared for the transmission of the data or state byte. The system controller must then address one or more devices as listeners, to allow the start of the data transfer cycle.

If the talker and the listener(s) are assigned, then the controller sends the ATN message with false value. Then, if the SPMS state is inactive, the talker function goes to the TACS state and the device can send messages to the listener device(s) of the system via the interface signal lines. The content of the messages is determined by the device functions.

The interface function remains in this active talker state until the controller sends the ATN message again, followed by an OTA message indicating that another talker's assignment will follow. The talker function then first returns to the TADS state, then, because of the OTA message, to the basic state. This organization of the talker function precludes the possibility of more than one active talker present in the system.

Apart from the data transfer, another task is possible for the talker function, provided that the device was designed to be capable of transmitting the Service Request (SRQ) message. Then it has to send an appropriate state byte to the bus if this is requested by the controller during serial poll.

The second part of the talker state diagram serves the purpose of the service request, consisting of the SPIS and the SPMS states. In normal circumstances the talker function is in the Serial Poll Idle State (SPIS). The function must go to the SPMS state after receiving the SPE message.

Following this, provided that the given device receives its own talker address and the ATN message then becomes false, the talker function moves from the TADS state to the SPAS state (*see* the top right-hand side of Figure 2.12).

In this state the talker function sends an appropriate state byte to the controller, indicating whether or not service request was made. In addition, it is possible to indicate what particular circumstance or state justifies the use of the service request.

The IEC recommendation sets down precisely the kinds of message that can be sent by the interface functions in their various states. For example, the STB message, indicating the state of the device, can be sent only in the SPAS state. Databyte (DAB) gets access to the bus in the TACS state only. The conditions for the transmission of uniline messages are defined by the recommendation in the same manner. For example, the END message, indicating the end of data transmission, can be sent on the EOI line in the TACS state only.

During the operation of the system the complete interface functions can fulfil several roles. The development of all roles or variants is not necessary in every case. For this reason the possible variants and their capabilities will be listed in the description of all interface functions. The possible omission of variants only limits the capability of the given instrument but has no effect on the operation of the complete system.

The permitted variants of the talker function can provide the following capabilities:

- *Basic talker*, permitting the transfer of data by the device to another device.
- *Serial poll*, permitting the transfer of a data byte to the controller to identify the device requesting service.

- *Talk only*, permitting the operation of a device as a talker in a system without a controller.
- *Unaddress (disable)*, if MLA, preventing a device possessing talker and listener functions receiving its own transmission.

The state diagram of the extended talker function (TE) is shown in Figure 2.13 and the names of the various states are listed in Table 2.5.

This function is differentiated from the talker function described above by the states relating to the secondary addressing. One such state is the TPIS, when the talker cannot recognize its own secondary address because it has not yet received a primary address. Remote messages cannot be sent in this state. If the MTA message is true and the ACDS state is active, then the extended talker function goes to the TPAS state. In this state the TE function can recognize its own secondary address.

The permitted variants of the extended talker function are identical to those of the talker function, supplemented by the secondary addressing.

The requirements concerning the talker and extended talker functions are the following:

- In the devices possessing these functions it must be ensured that the talker address (MTA) or the secondary address (MSA) can be altered by the user, and
- Every device possessing the talk only variant must be equipped with a manually operated switch capable of generating the ton message.

Table 2.5 The states of the extended talker function (TE)

Abbreviation	*State*
SPAS	serial poll active
SPIS	serial poll idle
SPMS	serial poll mode
TACS	talker active
TADS	talker addressed
TIDS	talker idle
TPAS	talker primary addressed
TPIS	talker primary idle

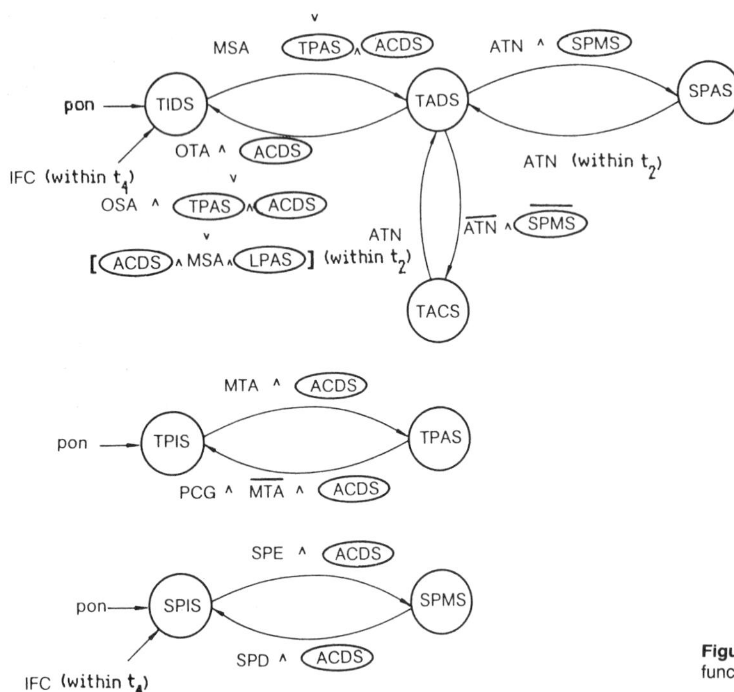

Figure 2.13 State diagram of the extended talker function

2.5.2.4 *Listener (L) and extended listener (LE) functions*

A device possessing listener function can receive dependent data from other devices via the interface. This capability exists only if the given function receives its own address from the control unit.

There are two variants of the listener function, differing only in their addressing modes. The normal function can be addressed with a 1-byte address, while the address of the extended listener function consist of two bytes. The capabilities of the two variants are identical in every other respect. In any device only one of the two listener functions can be realized.

The state diagram of the listener function is shown in Figure 2.14 and the names of the individual states are listed in Table 2.6.

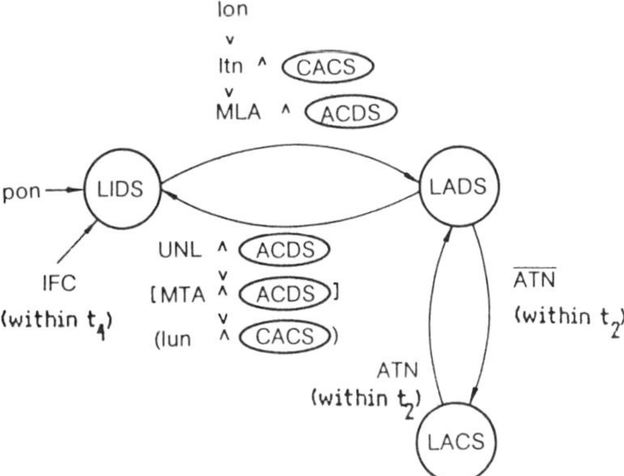

Figure 2.14 State diagram of the listener function

The start state of the function is the LIDS, if the power supply has been switched on (pon), or if an Interface Clear (IFC) has arrived. If the IFC message is false and the device switch generating the listener only (lon) message is being switched on, the listener function will be in the Listener Addressed State (LADS). The same process will take place if the device receives its own listener address in the start state and the receiver handshake interface function is in the ACDS state at the same time.

In the LADS state the listener function is prepared to receive device-dependent messages, but does not receive them yet. The listener function remains in this state until the system controller finishes addressing the units, the talker and the listener(s) participating in the information transfer. During this time the MTA and the MLA address designating messages are transmitted and the ATN message is true.

If the ATN message is false, the listener function transfers to the LACS state. This transfer must end in 200 nseconds, commencing from the ATN change. This upper time limit ensures that the listener is ready to receive data arriving from the talker.

In the LACS state the listener function can receive device-dependent data from the interface bus and can transmit them to the device functions.

Table 2.6 The states of the listener function (L)

Abbreviation	*State*
LACS	listener active
LADS	listener addressed
LIDS	listener idle

The listener function remains in this state until the controller changes the ATN message to true when the function goes back to the LADS state. Following this, it reverts to the start state as a result of the UNL message sent by the controller.

There is also an optional transfer from the LADS state to the start state. This will happen if the device addressed as listener receives its own talker address (MTA). This transfer prevents the reception of its own transmission, because the listener function goes to the start state instead of the LACS state after the addressing cycle.

The possible variants of the listener function can provide the following capabilities:

- *Basic listener*, allowing a device to receive device-dependent data from other devices.
- *Listen only*, allowing a device to operate as listener in a controller-less system.
- *Unaddress (disable)* if MTA, preventing a device from receiving its own transmission (devices possessing both listener and talker functions).

The state diagram of the extended listener is shown in Figure 2.15 and the names of the individual states are listed in Table 2.7.

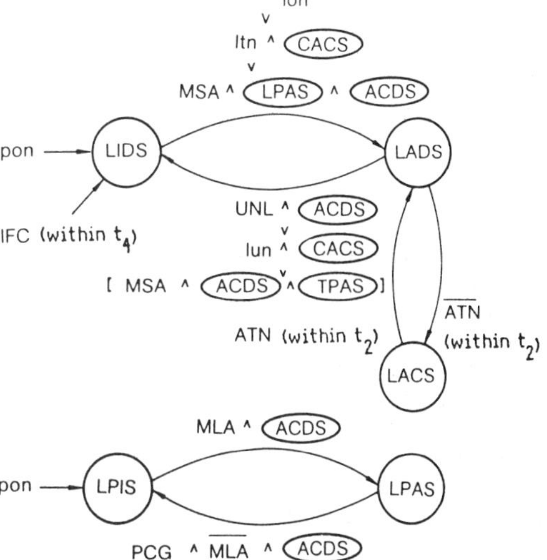

Figure 2.15 State diagram of the extended listener function

This function differs from the listener function discussed above essentially in the states concerning the secondary addressing. One of these states is the LPIS, in which the function can recognize its own primary address, but cannot react to its own secondary address. From this state following the switched-on state, the function goes to the LPAS state after having recognized the primary address (MLA). In this state the function can now recognize its own secondary address and can react to it.

Table 2.7 The states of the extended listener function (LE)

Abbreviation	*State*
LACS	listener active
LADS	listener addressed
LIDS	listener idle
LPAS	listener primary addressed
LPIS	listener primary idle

The possible variants of the extended listener function are identical to those of the listener function, augmented by the possibilities provided by the secondary addressing.

The requirements concerning the listener and the extended listener functions are:

- In the devices containing these functions it must be ensured that the listener address (MLA) or the secondary address (MSA) is user variable.
- Every device containing the listener only variant must be equipped with a manually-operated switch capable of generating the lon (listener on) message.

2.5.2.5 *Source handshake (SH) function*

This function ensures the capability to transfer multiline messages in the devices. The transfer of every individual multiline message is ensured by a transfer cycle between one source side and several receptor side handshake functions. The SH function controls the beginning and the end of the multiline message byte transmission via the DAV signal line.

The state diagram of the SH function is shown in Figure 2.16, and Table 2.8 lists the names of the individual states.

Whilst there is only one variant of the SH function, the Figure shows the entire structure. Without this the device cannot participate in the information exchange.

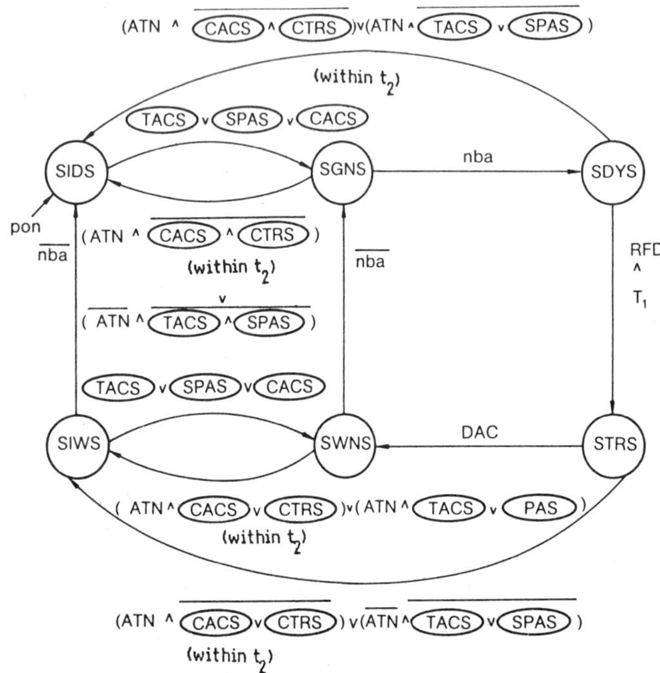

Figure 2.16 State diagram of the handshake function

Table 2.8 The states of the source handshake function (SH)

Abbreviation	*State*
SDYS	source delay
SGNS	source generate
SIDS	source idle
SIWS	source idle wait
STRS	source transfer
SWNS	source wait for new cycle

Table 2.9 The states of the acceptor handshake function (AH)

Abbreviation	*State*
ACDS	accept data
ACRS	acceptor ready
AIDS	acceptor idle
ANRS	acceptor not ready
AWNS	acceptor wait for new cycle

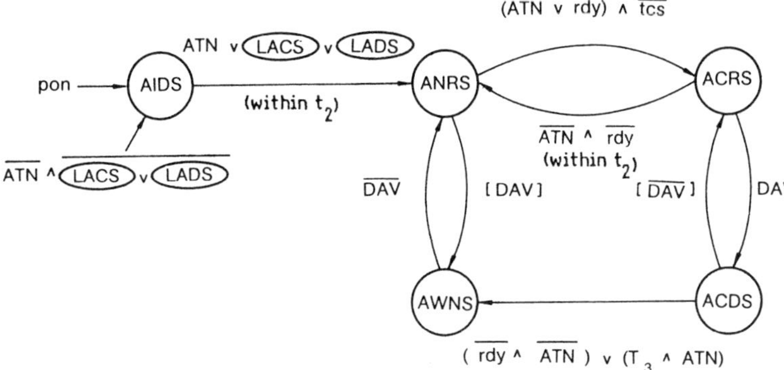

Figure 2.17 State diagram of the acceptor handshake function

2.5.2.6 *Acceptor handshake (AH) function*

The AH function ensures that the device can receive multiline messages. The transmission of every single message is ensured by a transfer cycle. The role of the AH function in this asynchronous data transfer is to communicate to the talker via the RFD signal line that it is ready to receive the new data and, via the DAC line, that it requires no more data.

The state diagram of the AH function is shown in Figure 2.17, and the names of the states are listed in Table 2.9.

Again there is only one usable version of the AH function, although the complete structure is shown in the Figure. Without it the device cannot accept data.

2.5.2.7 *Service request (SR) function*

A device possessing the SR function can ask for attention from the system controller in an asynchronous manner. One SR function allows service requests for one reason only. If a device can request service for several reasons, then a separate SR function must be used for each reason.

The state diagram of the SR function is shown in Figure 2.18, and the function names are listed in Table 2.10.

There is only one version of the SR function that provides the entire capability and without it the device is unsuitable for service request.

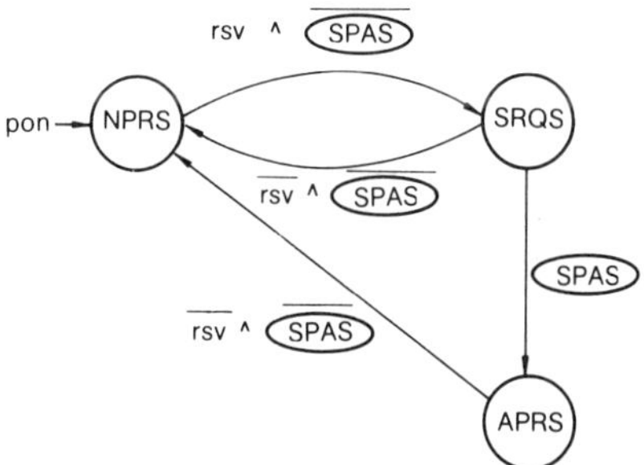

Figure 2.18 State diagram of the service request function

Table 2.10 The states of the service request function (SR)

Abbreviation	*State*
APRS	affirmative poll response
NPRS	negative poll response
SRQS	service request

2.5.2.8 *Remote/local (RL) function*

The device equipped with the RL function is capable of distinguishing between two different information sources. This function indicates to the device whether it should accept information originating from its front panel operational controls (local), or those arriving from the interface (remote). In remote control the device receives program data in its addressed listener function.

The state diagram of the RL function is shown in Figure 2.19, and the names of the function are listed in Table 2.11.

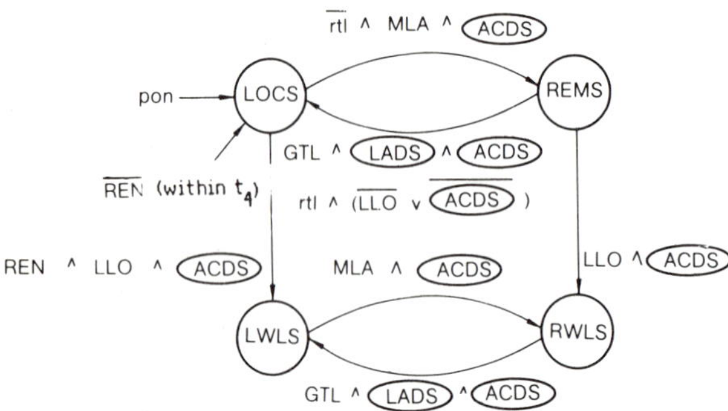

Figure 2.19 State diagram of the local/remote function

The permitted variants of the RL function can provide the following capabilities:

- *Basic remote/local*, ensuring that the control of the device can be changed from local to remote and from remote to local.
- *No local lock out*, this variant disallows the lock out of the 'return to local' (rtl) message.

Table 2.11 The states of the remote/local function (RL)

Abbreviation	*State*
LOCS	local control
LWLS	local with lockout
REMS	remote control
RWLS	remote with lockout

2.5.2.9 *Parallel poll (PP) function*

The parallel poll interface function allows a device to send a one-bit state message to the controller without having received a talker address previously. In parallel poll the signal lines of the data bus (DIO1 . . . DIO8) are used for the transfer of the state bits of the individual devices and therefore eight devices can participate simultaneously in the parallel poll. The assignment of the DIO signal lines to the individual devices must be carried out prior to the parallel poll cycle. More than one device can share a signal line, when the state information of the individual device goes onto the given signal line in conjunction with logic operators (AND, OR).

The organized intermittent execution of the parallel poll function is the task of the system controller. The state diagram is shown in Figure 2.20, and the names of the function states are listed in Table 2.12.

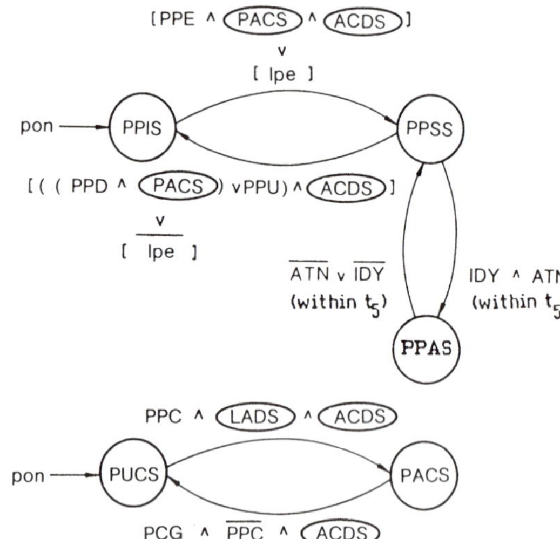

Figure 2.20 State diagram of the parallel poll function

The permitted variants of the parallel poll function can provide the following capabilities:

- *Parallel poll remote configuration*, allowing a device to communicate its state in a parallel poll cycle to the controller on the DIO bus line that was assigned by the PPC message.
- *Parallel poll local configuration*, allowing local configuration in the parallel poll.

Table 2.12 The states of the parallel poll function (PP)

Abbreviation	*State*
PACS	parallel poll addressed to configure
PPAS	parallel poll active
PPIS	parallel poll idle
PPSS	parallel poll standby
PUCS	parallel poll unaddressed to configure

2.5.2.10 *Device clear (DC) function*

The DC function permits the clearing of devices (their re-instatement to the starting state) one-by-one, or in groups. The group can be a subset of a system, or all the addressed devices in it.

The state diagram of the DC function is shown in Figure 2.21, and the names of the function states are listed in Table 2.13.

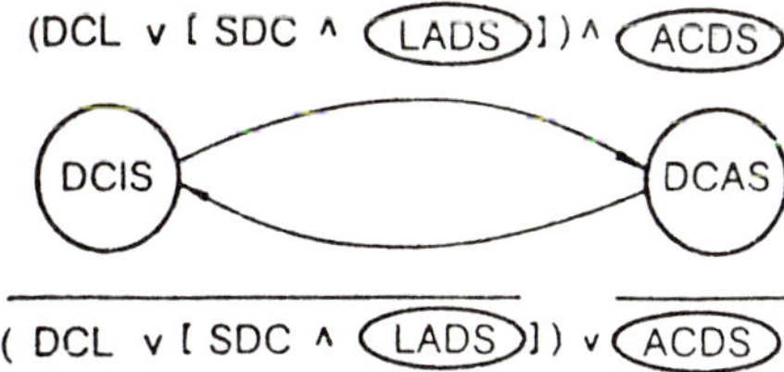

Figure 2.21 State diagram of the device clear function

The permitted variants of the device clear function can provide the following capabilities:

- *Complete device clear*, allowing the selective or simultaneous clearing of the system devices.
- *No selective device clear*, allowing only simultaneous clearing of all devices of the system.

Table 2.13 The states of the device clear function (DC)

Abbreviation	*State*
DCAS	device clear active
DCIS	device clear idle

2.5.2.11 *Device trigger (DT) function*

The device trigger function ensures the start of the basic operation process (e.g. measurement) of the devices from the interface, either one-by-one or in groups.

The state diagram of the device trigger function is shown in Figure 2.22, and the names of the function states are listed in Table 2.14.

The only usable variant of the device trigger function is the complete structure shown in the Figure.

2.5.2.12 *Controller (C) function*

The device possessing the controller function can send device addresses, universal commands and addressed commands to other devices via the interface. Apart from this, the controller function allows the device to carry out parallel poll to identify the device or devices requesting service, the clearing of the interface with the IFC message and the enabling of the devices for remote control with the REN signal.

The state diagram of the controller function is shown in Figure 2.23, and the names of the function states are listed in Table 2.15.

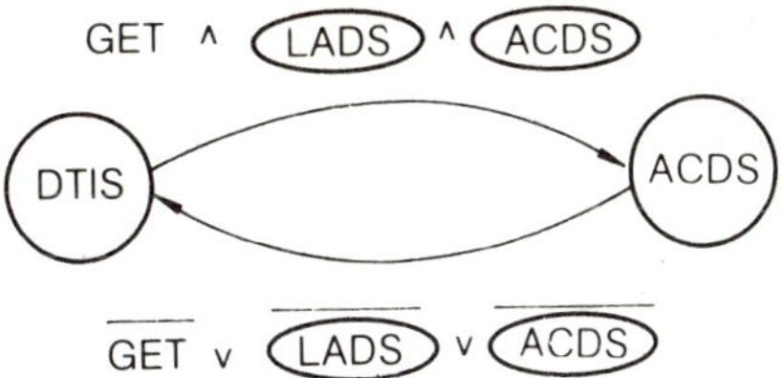

Figure 2.22 State diagram of the device trigger function

Table 2.14 The states of the device trigger function (DT)

Abbreviation	*State*
DTAS	device trigger active
DTIS	device trigger idle

Table 2.15 The states of the controller function (C)

Abbreviation	*State*
CACS	controller active
CADS	controller addressed
CAWS	controller active wait
CIDS	controller idle
CPPS	controller parallel poll
CPWS	controller parallel poll wait
CSBS	controller standby
CSNS	controller service not requested
CSRS	conroller service requested
CSWS	controller synchronous wait
CTRS	control transfer
SACS	system control active
SIAS	system control interface clear active
SIIS	system control interface clear idle
SINS	system control interface clear inactive
SNAS	system control not active
SRAS	system control remote enable active
SRIS	system control remote enable idle
SRNS	system control remote enable inactive

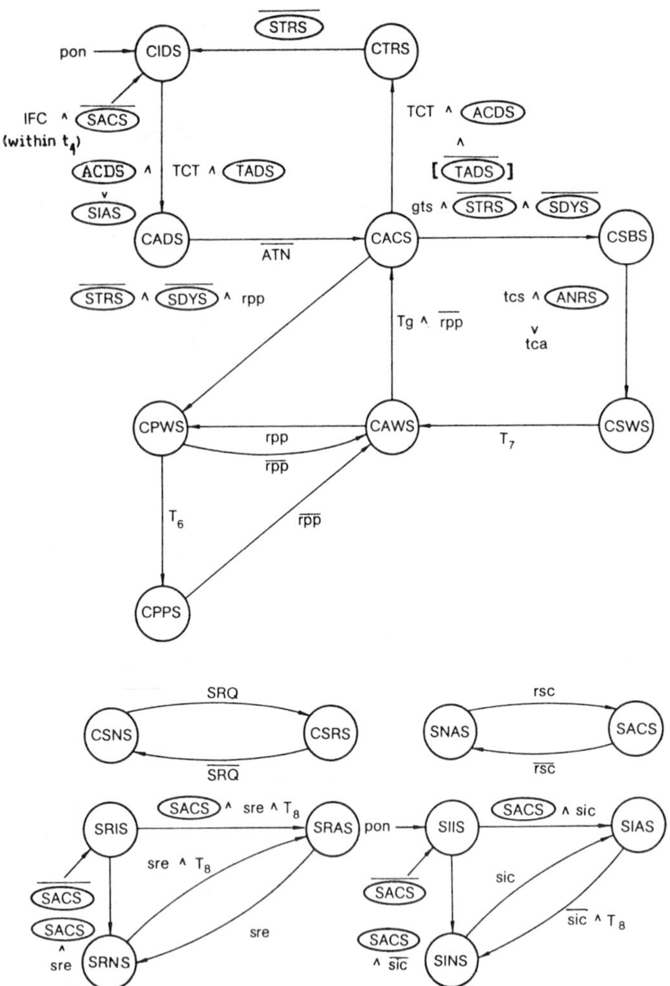

Figure 2.23 State diagram of the controller function

The permitted variants of the controller function can provide the following capabilities:

- *System controller*, allowing the device to transmit IFC and REN messages to the interface.
- *Send IFC and take charge*, allowing the transmission of the IFC message permitting the device to take over the controller role from another device.
- *Send REN*, allowing the controller to reset the device to remote.
- *Response to SQR*, allowing the controller to start a serial poll cycle after service request.
- *Send interface messages*, allowing the controller to send multiline messages to the interface signal lines.
- *Receive control*, allowing the controller to take over control from another device.
- *Pass control*, allowing the controller to give up control to another device.
- *Pass control to self*, allowing the controller to resume control from the controller on duty.
- *Parallel poll*, allowing the controller to execute a parallel poll.
- *Take control synchronously*, allowing the controller to take over control without detrimental effects on the data transfer in process.

The requirements and directives concerning the controller function are the following:

- If more than one device of a system possesses the controller function then, with the exception of one of them, all must be in the Controller Idle State (CIDS) permanently. The device containing the non-CIDS state is called the 'duty controller' of the system; and

- In any one system only one device's controller interface function may be in the System Active Control State (SACS), remaining there during the entire operation of the system, thus allowing the transmission of the IFC and REN messages at any time. This device is called the 'system controller'.

2.6 Message coding

In the IEC interface system the information transfer between the individual devices takes place in the form of remote messages. These remote messages – or interface messages – are those that are transmitted by the interface functions on the interface signal lines; device-dependent messages are those that determine the device functions.

There are fundamental differences between the two types of message. The interface messages are interface-dependent and their coding is defined precisely, in order to ensure compatibility of the devices. On the other hand, the device-dependent messages are system- and device-dependent and therefore a uniform coding system and message format cannot be standardized for them. However, a coding and format recommendation can be given even for the device-dependent messages and the adherence to such recommendations facilitates the design process of measuring system construction.

Figure 2.24 illustrates the relationship of the EC system functions and messages.

2.6.1 Coding of the interface messages

Table 2.16 contains the coding of the interface messages of the interface system. There are two types of message listed in the Table:

- messages characterized by the logic state of a single signal line, called 'Uniline messages' (U) (e.g. ATN), and
- messages characterized by the combined logic state of several signal lines, called 'Multiline messages (M) (e.g. DCL).

Some interface messages are defined as the logic combinations of other messages (e.g. OTA).

The interface messages can be classified according to their function thus:

- addressed command AC
- talker or listener address AD
- device dependent message DD
- handshake HS
- universal command UC
- secondary address SE
- state message ST

The Table contains the symbols of all message types and classes. It defines the necessary coding for the message transmission in the remote message rows. As the coding of the transmitted messages is the same, this also acts as a decoder for message reception. The meaning of the values in the Table are:

0 logic zero (level H)
1 logic one (level L)
X indifferent, as far as received message coding is concerned (driving is not allowed for the coding of transmitted data, unless directed by another message)
Y indifferent as far as transmitted message coding is concerned.

The value of uniline messages can be regarded as valid after the recognition of the assigned logic value. The validity of multiline messages is more complex than this. To avoid problems due to time lapses, these messages are correlated to the

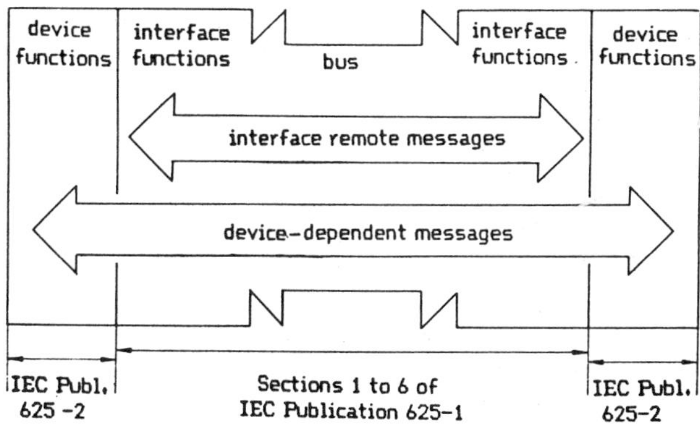

Figure 2.24 The relationship of the IEC interface functions and messages

Table 2.16 Coding of the IEC interface messages

Abbreviation	*Bus signal line mode*																*Message*	
	DIO8	DIO7	DIO6	DIO5	DIO4	DIO3	DIO2	DIO1	DAV	NRFD	NDAC	ATN	EOI	SRQ	IFC	REN	Type	Class
ACG	Y	0	0	0	X	X	X	X	X	X	X	1	X	X	X	X	M	AC
ATN	X	X	X	X	X	X	X	X	X	X	X	1	X	X	X	X	U	UC
DAB	D8	D7	D6	D5	D4	D3	D2	D1	X	X	X	0	X	X	X	X	M	DD
DAC	X	X	X	X	X	X	X	X	1	X	X	X	X	X	X	X	U	HS
DCL	Y	0	0	1	0	1	0	0	X	X	0	1	X	X	X	X	M	UC
END	X	X	X	X	X	X	X	X	X	X	X	0	1	X	X	X	U	ST
EOS	E8	E7	E6	E5	E4	E3	E2	E1	X	X	X	0	X	X	X	X	M	DD
GET	Y	0	0	0	1	0	0	0	X	X	X	1	X	X	X	2	M	AC
GTL	Y	0	0	0	0	0	0	1	X	X	X	1	X	X	X	X	M	AC
IDY	X	X	X	X	X	X	X	X	X	X	X	1	1	X	X	X	M	UC
IFC	X	X	X	X	X	X	X	X	X	X	X	X	X	X	1	X	U	U
LAG	Y	0	1	X	X	X	X	X	X	X	X	1	X	X	X	X	M	AD
LLO	Y	0	0	1	0	0	0	1	X	X	X	1	X	X	X	X	M	UC
MLA	Y	0	1	L5	L4	L3	L2	L1	X	X	X	1	X	X	X	X	M	AD
MTA	Y	1	0	T5	T4	T3	T2	T1	X	X	X	1	X	X	X	X	M	AD
MSA	Y	1	1	S5	S4	S3	S2	S1	X	X	X	1	X	X	X	X	M	SE
NUL	0	0	0	0	0	0	0	0	X	X	X	X	X	X	X	X	M	DD
OSA	(OSA=SCG∧MSA)																M	SE
OTA	(OTA=TAG∧MTA)																M	AD
PCG	(PCG=ACG∨UCG∨LAG∨TAG)																M	–
PPC	Y	0	0	0	0	1	0	1	X	X	X	1	X	X	X	X	M	AC
PPE	Y	1	1	0	P4	P3	P2	P1	X	X	X	1	X	X	X	X	M	SE
PPD	Y	1	1	1	D4	D3	D2	D1	X	X	X	1	X	X	X	X	M	SE
PPR1	X	X	X	X	X	X	X	1	X	X	X	1	1	X	X	X	U	ST
PPR2	X	X	X	X	X	X	1	X	X	X	X	1	1	X	X	X	U	ST
PPR3	X	X	X	X	X	1	X	X	X	X	X	1	1	X	X	X	U	ST
PPR4	X	X	X	X	1	X	X	X	X	X	X	1	1	X	X	X	U	ST
PPR5	X	X	X	1	X	X	X	X	X	X	X	1	1	X	X	X	U	ST
PPR6	X	X	1	X	X	X	X	X	X	X	X	1	1	X	X	X	U	ST
PPR7	X	1	X	X	X	X	X	X	X	X	X	1	1	X	X	X	U	ST
PPR8	1	X	X	X	X	X	X	X	X	X	X	1	1	X	X	X	U	ST
PPU	Y	0	0	1	0	1	0	1	X	X	X	1	X	X	X	X	M	UC
REN	X	X	X	X	X	X	X	X	X	X	X	X	X	X	X	1	U	UC
RFD	X	X	X	X	X	X	X	X	X	0	X	X	X	X	X	X	U	HS
RQS	X	1	X	X	X	X	X	X	X	X	X	0	X	X	X	X	U	ST
SCG	Y	1	1	X	X	X	X	X	X	X	X	1	X	X	X	X	M	SE
SDC	Y	0	0	0	0	1	0	0	X	X	X	1	X	X	X	X	M	AC
SPD	Y	0	0	1	1	0	0	1	X	X	X	1	X	X	X	X	M	UC
SPE	Y	0	0	1	1	0	0	0	X	X	X	1	X	X	X	X	M	UC
SRQ	X	X	X	X	X	X	X	X	X	X	X	X	X	1	X	X	U	ST
STB	S8	S7	S6	S5	S4	S3	S2	S1	X	X	X	0	Y	X	X	X	M	ST
TCT	Y	0	0	0	1	0	0	1	X	X	X	1	X	X	X	X	M	AC
TAG	Y	1	0	X	X	X	X	X	X	X	X	1	X	X	X	X	M	AD
UGG	Y	0	0	1	X	X	X	X	X	X	X	1	X	X	X	X	M	UC
UNL	Y	0	1	1	1	1	1	1	X	X	X	1	X	X	X	X	M	AD

handshake functions as well. A transmitted multiline message is valid until the source handshake function is in the Source Transmission State (STRS). A received multiline message is valid until the acceptor handshake function is in the data transfer state (ACDS).

2.6.2 Coding of device-dependent messages

The role of the IEC interface system is the transmission of device-dependent messages from device function to device function via the interface functions. The devices connected by the interface system receive and transmit various device-dependent messages at various times for various purposes.

The device-dependent messages can be classified as:

(a) Measurement data (e.g. characteristics measured by an instrument).
(b) Program or control data (e.g. setting up of the measurement functions of an instrument).
(c) State data (e.g. internal state of an instrument).
(d) Data to be displayed (e.g. alphanumeric messages).

The formats of the above messages vary even within the same measuring system; therefore in the emphatically general purpose IEC system – because of widely-varying applications and the different types of device used – it is neither possible nor even necessary, to provide the same format for the various device-dependent messages. Using the IEC interface, measuring systems requiring recommendations for the lowest level of the transmitted messages can be assembled, but there are also systems where this is necessary only at the highest level of data transmission.

The device-dependent messages are generated, processed or evaluated by the individual devices. Hence, the characteristics of these messages are strongly influenced by the type of system that will incorporate the device in question.

The converse of this is also true. The codes and formats used in a device will determine what systems are suitable for the device to operate efficiently. Accordingly, to ensure satisfactory system operation, the coding and format definitions of the individual devices must be carefully studied in the first stage of design.

2.6.2.1 The recommended message format

The device-dependent message represents an information parcel that has either a defined, or an implied, beginning and end. This information parcel is therefore generated, transmitted and evaluated or processed as a self-contained unit. The device-dependent messages contain an introductory or header (alpha) section, a main section, and a conclusion or ending (delimiter) section.

The message can be subdivided into data fields that characterize recognizable and separable data sections within the message. They are essential for optional data fields. Not all messages have to contain all possible data fields, but there are some that must be part of every message. Table 2.17 gives the symbols and the contents of each data field. The data fields denoted with an 'O' are optional; those denoted with a 'C' are compulsory; i.e. every message must contain them.

2.6.2.2 Coding of measurement data

The instruments usually transmit data as a result of the execution of a measurement task. The data is transferred to the bus while the instrument's talker interface function is in the TACS state. The length of the data output format can be varied for different type instruments and can contain both alphabetical and numerical information. The task of the former is to identify and explain the numerical part of the data output.

The representation of measurement data can contain five different data fields, each serving a different purpose. Of the data fields listed in Table 2.17, the use of X, Y and Z means mutual exclusion, since only one of these can be used in one message.

All data field are optional, except the V field, containing the numerical value. Due to the varied properties of the devices, the length of T, V and W

Table 2.17 Structure of the device-dependent messages

Message part	*Symbol*	*Contents*	*Character*
Introductory part (alpha)	T	definition of the type and quantity of the data	C
Main part (numeric)	U	sign or polarity of the data	O
	V	numeric value of the data	C
	W	sign of exponent	O
Conclusion part (delimiter)	X	string delimiter	O
	Y	block delimiter	O
	Z	record delimiter	O

fields can vary as required to describe and identify the numeric portion of the data string.

The contents of field T: Field T defines the type and the quality of the measurement data. It is generally recommended that this field should be as short as possible. The evaluation of data is made easier if the length of the field T is fixed for each device. If field T relates to measurement units, it is preferred to use basic units (F, V, A, m). Scaled values can be used as well, in which case the use of 10^3 and 10^{-3} multiples is preferred. Field T can contain commands describing the quality of the data following field V, such as overflow, calibration value, etc.

The contents of field U: Field U can contain a single character only. This field can be used to indicate the sign (polarity) of the data following in field V and is called 'signed representation'. If the sign is positive, or there is no need for the symbol, then field U can contain $\triangle$ (or blank space) and is called 'unsigned representation'.

The contents of field V: The length of field V, containing the numeric value of the measurement data, can be varied to satisfy the individual device requirements. In this field the most significant digit will be transmitted first. This is the only compulsory field for a message containing measurement data.

The contents of field W: Field W serves the purpose of giving the exponent part of the numeric value following in field V. This field consists of three compulsory parts in the following order:

(1) A single E denotes that the numbers following, including their sign, are to be interpreted as exponents of 10.
(2) A single + or − character denotes the sign of the exponent; if the exponent is zero, the + sign shall be used.
(3) The numeric value of the exponent is denoted by one or more numbers; the use of a two-digit number is preferred.

The contents of fields X, Y and Z: Detailed description of the contents and practical uses of these so-called 'delimiter fields' will be given in Section 2.6.2.7, 'Separation of message units', page 42.

2.6.2.3 *Program data*

Program data (e.g. measuring limits, operation mode, etc) are received by the measuring instruments in order to prepare for the execution of the measurement by the instrument whilst it is in the LACS state of the listener interface functions.

The messages containing program data can be assembled from the T and V data fields. Use of field T is compulsory, while the use of field V is optional.

The device-dependent requirements and capabilities are so widespread that it is not possible to prescribe the format of the program data. A few main directives will however, reduce the danger of possible ambiguities. All messages must contain at least one introductory character in field T to designate the functional use of the message. Appropriate selection of T field characters allows reduction or complete elimination of the number of separating delimiters between each functional message. The only recommendation concerning the data of the optional V field is that, in the case of exponential notation, the use of E must be circumspect to avoid ambiguity.

2.6.2.4 *State data*

State data are sent by the devices with the STB message in the SPAS state of the talker interface function on the DIO1 . . . DIO6 and the DIO8 signal lines. The service request (SQR) message is sent on the DIO7 signal line simultaneously with this message. The role of this simultaneously-transmitted double message, or in other words 'combined state byte', is to provide critical state data for the system controller. It is interpreted as the logic OR of detailed state data belonging to the same category.

The detailed state data are the following:

(a) *Other than normal state*: The combined bit-format state of the equation relating to the abnormal operation of the devices is transmitted on the DIO6 signal line.
Characteristic applications:
- error states within the device functions
- reception of false program data
- incomplete or false measurement data
- delimiting value or alarm state.

(b) *Engaged state*: The DIO5 line is used to indicate the combined state relating to the ready or engaged status of the main device functions (for instance voltage measurement, generate analogue output signal, etc).
(c) *Further states*: The DIO1 . . . DIO4 signal lines can be used by the system designer to indicate further combined states, or to express the above states in further detail. Assignment of the individual codes is free for maximum flexibility.

The marker bit transmitted on the DIO8 signal line can be used to alter or to extend the device-dependent part of the STB message. The recommended coding assignment for the STB message is shown in Table 2.18.

Table 2.18 Coding of the state byte (STB) messages

Message / *Logic value*	*RQS*	*STB*			
	DIO7	*DIO8*	*DIO6*	*DIO5*	*DIO4 . . . DIO1*
1	service requested	extended	abnormal	engaged	XXXX
0	service not requested	not extended	normal	ready	XXXX

X = device-dependent code assignment

2.6.2.5 *Data-shift methods*

With some devices it can happen that the input data bytes should be interpreted differently at a particular time, or in the case of given sequences. For instance, a display device (e.g. X-Y plotter) must receive a command for setting up the appropriate operating conditions first and then receive the specific data of the points to be displayed. In such cases there is a need for methods that allow the diversion of data traffic from the programme data to the data for display. In other words, methods are needed that help the listener device to distinguish between the different quality data groups to be output.

Data-shift can be executed in two ways. To some extent multiple-listener addresses can designate program data and data for display. The other solution is to use one listener address and to have a device-dependent shift-code chosen by the designer from the code fields available. Naturally the assignment of a shift-code (or codes) reduces the size of code fields remaining for the program and display data.

2.6.2.6 *Representation of data*

The following forms of data may be contained in device-dependent remote messages:

- alphabetical character (e.g. A, B, e, etc);
- numeric character (e.g. decimal (0, 1, 2), binary (2's complement), etc);
- symbols (e.g. +, −, :, =, etc).

The IEC interface recommendation does not contain compulsory specifications for the coding of device-dependent remote messages but recommends the use of the 7-bit ISO code. Table 2.19 shows this internationally laid down code system. Coding of two data types, the alphanumeric characters and symbols are evident from the Table. There are several possibilities, however, in representation of numeric data.

Representation of decimal data: Using the ISO 7-bit code, there are three possible forms for the numeric representation of decimal data. the NR1, NR2 and NR3 numeric representation methods apply the characters below:

1 2 3 4 5 6 7 8 9 0 E + − . ,

The NR1, NR2 and NR3 numeric representation methods do not specify the number of digits in the represented data; therefore, in the following examples the number of digits does not imply that it is necessarily advisable to transmit that number of digits.

NR1 numeric representation: The NR1 implicit decimal representation of numeric values describes integers; the decimal point is interpreted at the end of the number string, but is not transmitted. The 'signed representation' and the 'unsigned representation' should both contain at least one digit; no other signs or final blank space (△) are allowed in the field. If the length of the field is fixed, however, there can be blank spaces at the beginning of the number string. The signed representation of ZERO numeric value shall have either a positive sign or a space.

Table 2.20 lists a few examples of the NR1 numeric representation, where the underlined examples are the recommended notation.

The NR1 representation format is advantageous for numeric data that can be generated or interpreted in a limited or fixed range; or for the transmission of a large amount of fixed format data between devices.

NR2 numeric representation: The NR2 method represents the numeric value explicitly with a decimal point. For clarity, at least one digit should precede the decimal sign in both 'signed representation' and 'unsigned representation'.

Table 2.21 shows a few examples of the NR2 numeric representation. It is important to note that if a comma is used as a decimal sign, then it cannot be used as a delimiter.

Table 2.19 Multiline interface message representations (ISO 7-bit or ASCII code)

Bits (2)				b_7 b_6 b_5	0 0 0	MSG (1)	0 0 1	MSG	0 1 0	MSG	0 1 1	MSG	1 0 0	MSG	1 0 1	MSG	1 1 0	MSG	1 1 1	MSG
b_4	b_3	b_2	b_1	column / row	0		1		2		3		4		5		6		7	
0	0	0	0	0	NUL		DLE		SP	MLA DEVICE LISTENER ADDRESSES	0	MLA DEVICE LISTENER ADDRESSES	•	MTA DEVICE TALKER ADDRESSES	P	MTA DEVICE TALKER ADDRESSES		GIVEN BY THE PCG CODE	p	GIVEN BY THE PCG CODE
0	0	0	1	1	SOH	GTL	DC1	LLO	!		1		A		Q		a		q	
0	0	1	0	2	STX		DC2		"		2		B		R		b		r	
0	0	1	1	3	ETX		DC3		#		3		C		S		c		s	
0	1	0	0	4	EOT	SDC	DC4	DCL	$		4		D		T		d		t	
0	1	0	1	5	ENQ	PPC (3)	NAK	PPU	%		5		E		U		e		u	
0	1	1	0	6	ACK		SYN		&		6		F		V		f		v	
0	1	1	1	7	BEL		ETB		'		7		G		W		g		w	
1	0	0	0	8	BS	GET	CAN	SPE	(		8		H		X		h		x	
1	0	0	1	9	HT	TCT	EM	SPD	)		9		I		Y		i		y	
1	0	1	0	10	LF		SUB		*		:		J		Z		j		z	
1	0	1	1	11	VT		ESC		+		;		K		(		k		{	
1	1	0	0	12	FF		FS		,		<		L		/		l		:	
1	1	0	1	13	CR		GS		–		=		M		)		m		}	
1	1	1	0	14	SO		RS		.		>		N		^		n		~	
1	1	1	1	15	SI		US		/	(4)	?	UNL	O		—	UNT	o		DEL	

ADDRESSED COMMAND GROUP — GENERAL COMMAND GROUP — LISTENER ADDRESS GROUP — TALKER ADDRESS GROUP → PRIMARY COMMAND GROUP

SECONDARY COMMAND GROUP

Notes (1) MSG = INTERFACE MESSAGE (3) REQUIRES SECONDARY COMMAND

Table 2.20 Examples of NR1 numeric representation

Normal format	*Unsigned NR1*	*Signed NR1*
4902	0004902	+004902
	△△04902	△+04902
	△△△4902	△△+4902
		△△△4902
+1234	0001234	+001234
	△△△1234	△△+1234
		△△△1234
−56780	cannot be represented	−056780
		△−56780
0	0000000	+000000
	△△△△△△0	△△△△△+0
		△△△△△△0

Table 2.21 Examples of NR2 numeric representation

Normal format	*Unsigned NR2*	*Signed NR2*
1327.	1327.000	+1327.00
	0001327.	△△+1237.
	△△△1327.	△△△1327.
123.45	00123.45	△+123.45
	△△123.45	△△123.45
1237.0	△△1237.0	△+1237.0
		△△1237.0
.00001	00.00001	+0.00001
−5.678	cannot be represented	−5.67800
		−05.6780
0	000.0000	+0.00000
	△△△△△0.0	△△△△+0.0
		△△△△△0.0
		△△△△△△0.

The NR2 representation method is advantageous for data generated and evaluated in a limited range, or if the transferred data is to be interpreted by the user.

NR3 numeric representation: The NR3 numeric representation consists of decimal signed and exponent signed representation. This representation method should be used if the range of the measuring instrument is large, or if the characteristic value range of the transmitted (and received) data cannot be predicted in advance.

Table 2.22 lists a few examples of the use of the NR3 representation method. The underlined examples denote the recommended notation.

Representation of non-decimal data: The ISO 7-bit code is suitable for the representation of non-decimal data, using appropriate value assignment.

Representation of binary numeric values: The binary numeric values can be represented with a

Table 2.22 Examples of NR3 numeric representation

Normal format	*First recommendation*	*Second recommendation*	*Third recommendation*
5600	005.6E+03	00.56E+04	000056E+02
	+05.6E+03	+0.56E+04	+0056E+02
	△△5.6E+03	△0.56E+04	△△△56E+02
	△+5.6E+03		△△+56E+02
0.00002	0020.E−06	000.2E−04	
	00020E−06	0.200E−04	00002E−05
	+0020E−06	+00.2E−04	+0002E−05
	△△△20E−06	△△0.2E−04	△△△△2E−05
	△△+20E−06	△+0.2E−04	△△△+2E−05
−4.2	−△04.2E+00	−0.42E+01	0042E−01
	△−4.2E+00		△△42E−01
0	00000E+00	0.00000E+00	00000E+00
	+0000E+00	+0.00E+00	+0000E+00
	△△△△0E+00	△△0.0E+00	△△△△0E+00
	△△△+0E+00	△+0.0E+00	△△△+0E+00

subset of the decimal number set. Field V can contain only the following characters in the case of binary numeric representation:

0 1

Representation of octal numeric values: Similar to the binary numeric values, the octal numeric values can be represented with a subset of the decimal number set. The only permitted character set of the octal representation is:

0 1 2 3 4 5 6 7 .

Hexadecimal numeric value representation: An extended character set is necessary for hexadecimal representation. The character set below is recommended in the IEC proposal:

0 1 2 3 4 5 6 7 8 9 A B C D E F .

Care must be taken when using this representation method that the A . . . F characters are not confused with the symbols of the introductory field (T).

2.6.2.7 *Separation of message units*

In the data traffic between the devices of the IEC interface system there are often correlating data sets that are found in pairs (e.g. amplitude and phase), or in other multiple connections. Data groups such as these are generated, processed and evaluated together, as higher order sets of the individual basic message units.

There are the following type of data formats:

- file,
- record,
- block,
- string.

In practice, separation of three basic levels, the string, the block and the record, becomes necessary. The so-called delimiters serve this purpose, by facilitating the evaluation of complex messages. Figure 2.25 shows the position of the delimiters in a complex message. It is evident from the Figure that two delimiters cannot be placed in adjacent positions, even when they are different types.

String delimiters, X: The smallest unit of the data format is the string, consisting of a series of characters or bytes constructing the data set and is handled by both the data source and the receiver as a unit. String delimiters are used to separate either variables (e.g. amplitude, phase), or series of identical type data, in the TACS state of the talker function of a given device.

Block delimiters Y: Block delimiters are used to conclude either the individual message unit of a single measured value or a group of related message units.

Record delimiters, Z: The record delimiter concludes a data block series or in special circumstances a single data block. The record delimiter is the highest order delimiter used in the IEC standard.

The use of the delimiters: The function of the information on the data lines of the IEC interface (DIO1 . . . DIO8) is determined by the state of the ATN message. In the true state of the ATN message there is a general address or command on the data bus, and hence there is no need for delimiters. In the false state of the ATN message there are device-dependent messages on the data bus and parts of these are separated by delimiters.

The string delimiters are used to separate data sent in series. It is characteristic of their use that the talker function is in the TACS state.

During the use of block delimiters the talker function of the device can transfer to the TADS or TIDS state from the TACS state, but can remain in the TACS state as well. For example, the use of the block delimiter is justified after a message unit originating from a single measurement, if it is likely that the device will transmit another message unit at a later time, without the intervention of the controller or another device.

Record delimiters are generally used when the talker function changes to the TIDS state. It is usual, although not compulsory, to reset the ATN

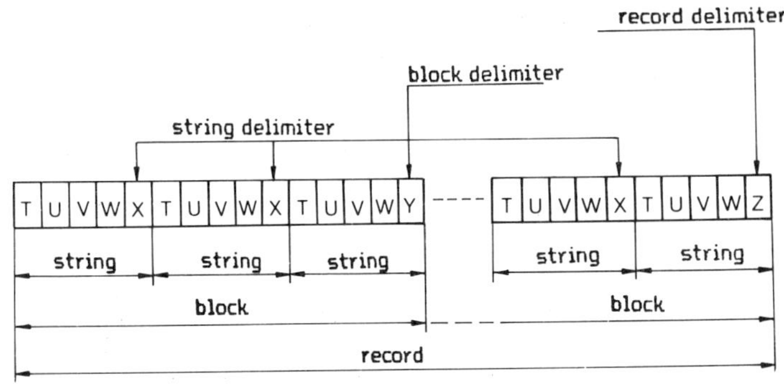

Figure 2.25 Position of the delimiters in a complex message series

Table 2.23 Recommended delimiter codes

Field	*Delimiter*	*Recommended ISO 7-bit code*	*Other code*
X	string	, or ;	
Y	block	ETB or CRLF	ETB Λ END
Z	record	ETX or END Λ DAB	ETX Λ END

message to 'true' after the record delimiter is used, to designate the new talker.

The controller function can interrupt a string, a block or a record containing multiple message units synchronously, in order to execute higher-priority tasks. In such cases it is advisable that the interupted device should start the data transfer at a later time with the byte following the byte transferred before the interruption.

2.6.2.8 *Magnitude assignment*

It is a general coding rule that in the IEC interface system the least significant bit of the code is assigned to the data line denoted by the lowest number. When an 8-bit binary code is used, the DIO1 . . . DIO8 lines transfer the $2^0 \ldots 2^7$ bits. In the case of a 5-bit code, the DIO1 . . . DIO5 lines transfer the $2^0 \ldots 2^4$ bits. Any unused signal lines must be set to passive (false) level.

Table 2.24 lists recommendations for the assignment of a few generally used code magnitudes. During the representation of the so-called 'packed numbers' shown in the Table, the most significant digit transfers to the DIO5 . . . DIO8 line. The X character used in the Table denotes either zero or a parity bit.

2.6.2.9 *Recommended SI units*

Tables 2.25, 2.26 and 2.27 contain the recommended unit symbols for use in the device-dependent messages of the IEC system. Table 2.28 lists the recommended multiplication factor symbols to represent the numeric value of data.

Table 2.24 Magnitude assignment for various codes

Code \ *Dio lines*	DIO8	DIO7	DIO6	DIO5	DIO4	DIO3	DIO2	DIO1
ISO 7-bit		b_7	b_6	b_5	b_4	b_3	b_2	b_1
Binary	2^7	2^6	2^5	2^4	2^3	2^2	2^1	2^0
Packed binary coded octal	X	2^2	2^1	2^0	X	2^2	2^1	2^0
Packed binary coded hexadecimal	2^3	2^2	2^1	2^0	2^3	2^2	2^1	2^0
Binary coded hexadecimal	X	0	0	0	2^3	2^2	2^1	2^0
Packed BCD	2^3	2^2	2^1	2^0	2^3	2^2	2^1	2^0
BCD	X	0	0	0	2^3	2^2	2^1	2^0

Table 2.23 lists the recommended notation for the string, block and record delimiters. There are a few important points concerning the Table:

(1) If a block contains only one string, then there is no need to use a string delimiter as well.
(2) If a record contains only one block, then there is no need to use a block delimiter as well.
(3) The ETB and ETX codes denote the state of the DIO lines set by the ISO code.
(4) The END message is transferred by the EOI line – if this message is used as a delimiter, then it should be transmitted simultaneously with the last data byte.
(5) If comma is used as a decimal sign, then it cannot be used as a delimiter.

Table 2.25 Basic SI unit symbols

Unit	*Standard symbol*	*Upper-case symbol*
metre	m	M
kilogram	kg	KG
second	s	S
ampere	A	A
kelvin	K	K
mole	mol	MOL
candela	cd	CD

Table 2.26 Derived SI unit symbols

Unit	*Standard symbol*	*Upper-case symbol*
hertz	Hz	HZ
newton	N	N
pascal	Pa	PA
joule	J	J
watt	W	W
coulomb	C	C
volt	V	V
farad	F	F
ohm	Ω	OHM
siemens	S	SIE
weber	Wb	WB
tesla	T	T
henry	H	H
lux	lx	LX
bel	B	B

Table 2.27 Miscellaneous symbols

Unit	*Standard symbol*	*Upper-case symbol*
decimal degree (grade, angle)	g (s)	GON
degree (angle)	° (s)	DEG
minute (angle)	′ (s)	MNT
second (angle)	″ (s)	SEC
litre	l	L
minute (time)	min	MIN
hour	h	HR
day	d	D
year	a	ANN
gram	g	G
tonne	t	TNE
bar	bar	BAR
poise	p	P
stokes	St	ST
electronvolt	eV	EV
degree Celsius	°C	CEL

2.7 The operation of the IEC interface system

Let us now examine with the help of a practical example the operation of the IEC interface system discussed in detail above.

In Figure 2.26 typical measuring system units are shown: the programmable desk calculator operating as controller, the programmable power supply connected to the unit under test, the function generator, the multimeter, the frequency meter and

Table 2.28 Multiplication factor symbols

Multiplication factor name	*Unit multiplication factor*	*International (standard) symbol*	*Upper-case symbol*
tera	10^{12}	T	T
giga	10^{9}	G	G
mega	10^{6}	M	MA
kilo	10^{3}	k	K
hecto	10^{2}	h	H
deca	10^{1}	da	DA
deci	10^{-1}	d	D
centi	10^{-2}	c	C
milli	10^{-3}	m	M
micro	10^{-6}	μ	U
nano	10^{-9}	n	N
pico	10^{-12}	p	P
femto	10^{-15}	f	F
atto	10^{-18}	a	A

the printer – whose task is the recording of measurement data – are all connected to the common interface bus.

The basic requirement for the system to operate is that all devices should possess an IEC interface unit and that those interfaces should contain the basic interface functions to enable the user to program and operate the individual devices efficiently.

The complete measuring system, as it is evident from the Figure, can be divided into two functional parts. The analogue-measuring system contains the device-dependent functions of the devices that are connected to the unit under test, while the digital interface system contains the interface functions and the bus system ensuring the connection.

2.7.1 Address assignment

The devices connected in the IEC interface system can be identified with the aid of the listener or talker addresses. The controller designates the active listeners or talkers by sending these addresses prior to the data transfer. A device can have a listener address, a talker address, or both. In the system shown in Figure 2.26 the multimeter, for example, has both listener and talker addresses. The instrument is designated to receive program data with the listener address, while the talker address calls for the output of measurement data. Conversely, the printer connected to the system has only a listener address, as this device only receives data.

The addresses are represented with a 7-bit binary code in the IEC system. The first 2 bits determine

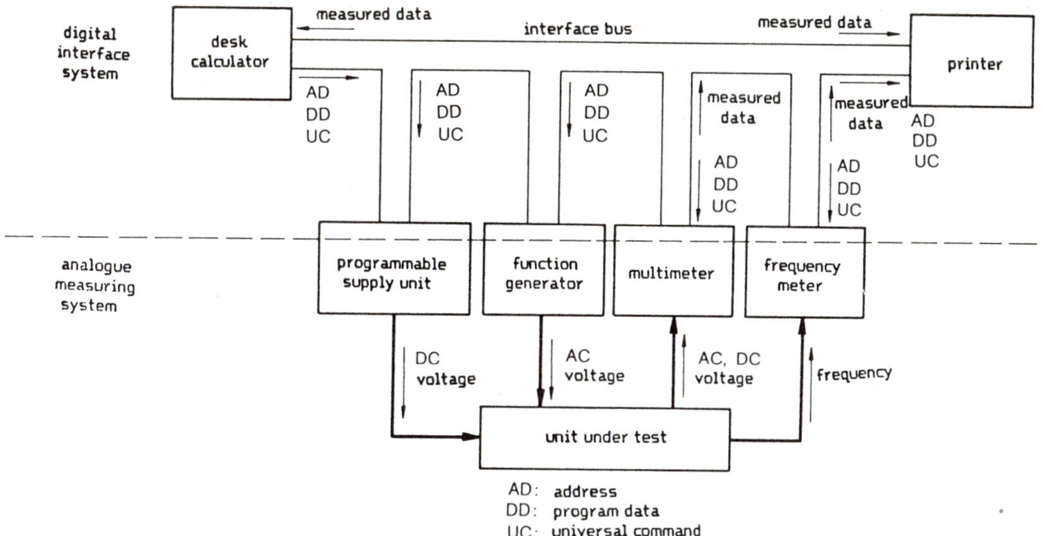

Figure 2.26 Structure of a typical IEC measuring system

the character of the address (listener or talker). The last 5 bits determine the individual addresses:

Magnitude of bits:	b_7 b_6 b_5 b_4 b_3 b_2 b_1
Talker address:	1 0 A_5 A_4 A_3 A_2 A_1
Listener address:	0 1

The rules of the interface system do not permit the use of addresses in which the value of the last 5 bits is 1 (1011111 or 0111111). Hence, 31 different talkers and listeners can be identified using primary addressing (1-byte-long address). The secondary address is determined by the $S_5 \ldots S_1$ bits of the 'My Secondary Address' (MSA) message code of the devices containing the extended listener or the extended talker function. A total of 961 listeners and 961 talkers can be distinguished with secondary addressing.

Assignment of the addresses of the devices connected in the measuring system is the task of the user. Care should be taken that any two devices should not normally have identical addresses. Two or more listeners can have identical addresses, however, provided that they are to receive the same information. Two talkers cannot in any circumstances have identical addresses.

In the case of the IEC-compatible instruments the assignment of the listener and talker addresses is generally achieved by switches or variable cross-connections located at the back of the instrument (Figure 2.27). In the operation manual of the instrument the positions of the switches and cross-connections and the mode of address assignment is described in detail. It should be borne in mind that one particular position of the address switches or cross-connections determines the two addresses, a talker address and a listener address.

Certain kinds of device have a state achieved by the use of a separate switch, by which they are addressable. In another position of the switch these devices are in the 'listen only' or 'talk only' operational mode, according to the nature of the device. This solution becomes important in systems not containing a controller.

Some devices can have two listener or two talker addresses. For example, the device receives the program data on one of its listener addresses and the measurement data on the other. Sometimes it is possible that an instrument transmits measurement data given in two different formats from two different addresses. On the back of these instruments there are only four address-assigning switches or cross-connections. Any state of these assigns two listener and two talker addresses.

It can happen that the manufacturers deliver certain devices with pre-set listener and talker addresses. For example, Hewlett-Packard Type 9830A calculators are marketed with a factory-set 'U' talker and '5' listener addresses. In such cases the user is somewhat limited in the address assignment, as these factory-set addresses can be changed only by internal rewiring.

Figure 2.27 Address selection on the back of a Hewlett-Packard instrument

2.7.2 Operation sequence

Let us assume that the execution of the measurement tasks of the system shown in Figure 2.26 takes place in two basic cycles:

I The calculator programs the instruments and starts the measurements, the resultant data being transmitted to the calculator for processing.

II The calculator programs the instruments and starts the measurements and the resultant measurement data are then transmitted to the printer.

The execution of each measurement cycle takes place in several steps in a pre-determined sequence:

Cycle I

(1) The calculator sets the interface system to the basic start state with the transmission of the 'Interface Clear' (IFC) message.

(2) The calculator sets the internal functions of the devices to a determined state with the transmission of the 'Device Clear' (DCL) message.

(3) The calculator first transmits the listener address of the programmable power supply, then transmits program data to this device.

(4) The calculator issues the 'Unlisten' (UNL) command followed by the listener address for the next device and sends program data to it.

(5) Step 4 is repeated until all devices participating in the measurement are brought to the programmed state. Then the 'Unlisten' (UNL) command is issued again.

(6) The calculator sends the listener address of the chosen instrument (e.g. the frequency meter), then issues the program code necessary to start the measurement.

(7) The calculator issues the 'Unlisten' (UNL) command, gives itself a listener address, then issues a talker address to the frequency meter.

(8) At the end of the frequency measurement the frequency meter sends the measurement data to the addressed listener (calculator).

Cycle II

(Steps 1 . . . 6 are identical to the appropriate steps of Cycle I).

(7) The calculator issues the 'Unlisten' (UNL) command, designates the printer as listener, then the frequency meter as talker.

(8) The frequency meter sends the measurement data to the designated listener (printer).

If the calculator addresses both itself and the printer as listeners, then the measurement data can be received by both devices. The handshake process controlling the data transfer ensures satisfactory operation even in cases such as this, where the data accepting speeds of the two devices are very different.

2.7.3 Serial poll

The devices of the measuring system can request service from the controller via the serial poll process. One reason for service request amongst others, can be the termination of the given measurement cycle. Returning to the example above, the multimeter is capable of measurement data output about 100 mseconds after the measurement command is received. Another reason can be the detection of some critical condition, e.g. overload of the power supply, lack of paper for the printer, or some other reason that in the design stage was deemed to be important and requiring attention.

The serial poll commences if one device (in the present case the multimeter) sends a true level to the 'Service Request' (SRQ) line. The controller (in our case the calculator) has been monitoring the state of the SRQ line and, after detecting the true level, begins to execute the serial poll, i.e. ascertains both the identity of the device requesting service and the reason for the request. The controller issues a true level onto the ATN line, then sends the Unlisten (UNL) and 'Serial Poll Enable' (SPE) commands followed by the talker address (TAD) of the first device. If the device addressed first was the one issuing the service request, then this device gives level 1 on the DIO7 line at the time the ATN message goes false. Simultaneously, the device sends the data byte indicating the reason for the service request to the rest of the data lines. If the device requesting the service was not the first addressed device, then the controller continues to address devices until it contacts the one that requested service. The serial poll process ends with the 'Serial Poll Disallowed' (SPD) message. The serial poll is illustrated on the flow diagram shown in Figure 2.28.

The serial poll is the most effective service request method. It allows devices at any time and in any state of the interface to request attention from the controller. However, it suffers from the disadvantage that it is a rather slow process, owing to the large number of messages exchanged.

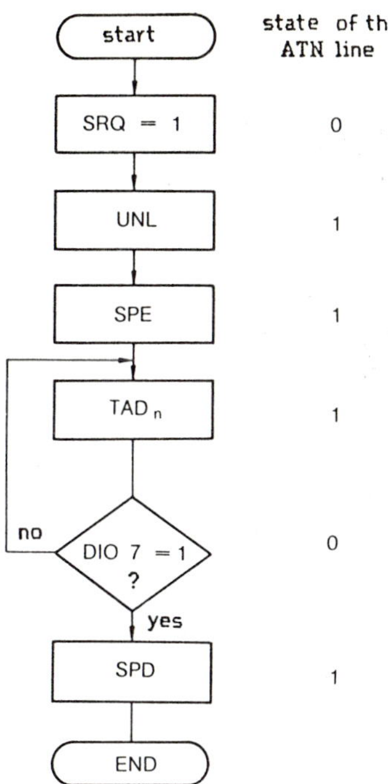

Figure 2.28 Parallel poll flow diagram

2.7.4 Parallel poll

The parallel poll is another method for service request. It is a faster process, but is bound by certain conditions.

This happens in such a way that the devices called upon by the controller indicate their service request needs on the designated line of the data bus. The assignment of the DIO lines to the individual devices must be carried out prior to the parallel poll cycle, either by a local message produced by a switch, or by a remote message sent by the controller. During the assignment the necessary logic value (0 or 1) must also be issued by the device to indicate service request.

If more than one device uses a signal line, then the value on the line can be the combination of the logic AND or the logic OR, depending on whether the devices use the 0 or the 1 value for service request.

Using the parallel poll the service requests can be transmitted only when the controller calls upon the devices to do so.

2.7.5 Local and remote control

The instruments connected in the IEC system can be controlled with either the switches or other handling devices located on their operators panel, or with the messages sent via the interface.

In the local control operation mode the instrument cannot send or receive messages to or from the bus.

The transfer to the remote control operation mode is carried out with the 'Remote Enabled' (REN) message. The controller switches an L level onto the REN line, then issues the listener addresses of the instruments designated for remote control one-by-one. The devices set to remote control in this way remain in this operation mode until the controller maintains level L on the REN line.

Apart from this, there are two ways of returning to the local control operation mode. The 'Go To Local' (GTL) addressed command resets all active listeners to local control. Individual resetting of devices to local control can be executed by the switch on the operators panel. Prohibition of this switch is possible by using the 'Local Lock Out' (LLO) message. With this message accidental switching over that could endanger the automatic measuring process can be prevented.

2.7.6 Device clear

The units of the automatic measuring system must be transferred to the start state after having been switched on. This can be executed by two interface messages. The 'Interface Clear' (IFC) is a uniline message, resetting the interface functions, while the 'Device Clear' (DCL) general command resets the device functions. With the 'Selected Device Clear' (SDC) addressed command the device functions of active listeners can be reset. This allows the device functions of a single device or device group to be reset.

2.7.7 Synchronization of measurements

In automatic measuring systems it can be a requirement that several instruments must execute measurement tasks simultaneously. Very often the commencement of measuring operations should coincide with the generation of a signal. For instance, in the examination of the response given to a single impulse, the impulse generator and the measuring instrument (eg. storage oscilloscope) must be started at the same time.

In IEC measuring systems this kind of synchronization is executed with the 'Group Execute Trigger' (GET) addressed command. This message starts all active listeners. The start can be executed with the following message sequence:

LAD 1	
LAD 4	(designation of listeners)
LAD 6	
GET	(start of listeners having the addresses 1, 4 and 6)
UNL	(preparation of further interface operations)

2.7.8 Message formats

The structure of the messages cannot be prescribed in a rigid and compulsory manner in a general purpose interface system, as it would limit the application areas. For this reason, the IEC interface standard recommends several message type variations for guidance of instrument designers and system users. Nevertheless, in special cases deviation from the recommendations becomes necessary. Despite this, it is very important to keep the directives in mind, as the selection of the recommended formats makes the programming tasks of the user much simpler.

2.7.8.1 Coding of measurement data

There are three different situations possible in the data traffic of the interface system. The data can be transferred:

(a) to the controller from a device,
(b) to one or more devices from the controller, and
(c) to one or more other devices from a device.

In all three cases the alpha characters as well as the numeric data must be transferred.

For example, let us assume that the multimeter measures +12 002 V direct voltage on the 10 V range and the measurement result is transmitted in NR3 (exponential) format after a single measurement. The alpha characters found in the introductory T field precede the numeric data in the U, V and W fields. The former denotes the operation mode (direct voltage, DC) and the quality of the measured value (OL, overload). The character string is concluded with the CRLF block delimiter, after one measured value is output:

OLDC+12002.E−03CRLF

If two identical character measurement values are transmitted, then a string delimiter can be used between the two values and a record delimiter to conclude the message. For example, if a frequency counter measures two values in a measurement series (4.23 MHz and 2.6 MHz), then this can be transmitted with the following message:

FMHZ4.23;FKHZ2,60NL

2.7.8.2 Coding of program data

Programmed control of an instrument is in essence the substitution of manual control with appropriately-coded digital signals. Hence, the manual-operation mode of the instrument must therefore be known for programming. In our example, the power supply acting as the power source of the unit under test can be programmed to 5.25 V with a 120 mA current limit like this:

V5,25E+00I120E−03

In the case of a 10 V measuring limit the multimeter can be programmed to measure direct voltage on receipt of an internal trigger, for example:

F0R4T1M3E

In this message:

- F0 is the measurement operation mode (function)
- R4 is the range
- T1 is the starting method (trigger)
- M3 is the data output mode
- E is the execution command.

Naturally, the content and the format of the messages transmitting the programming data of the devices are determined by the individual characteristics of the devices. The operation manuals of IEC instruments which are interface-compatible contain detailed guidelines to facilitate programming.

3 Controllers

The operation of an automatic measuring system is organized by the control units. As well as controlling the acquisition of signals and the operation of the instruments, it evaluates the measurement results, displays them in a format that is easily interpreted by the user, and sometimes processes the data. The capacity of automatic systems therefore depends greatly on the characteristics of the controller.

Due to the variety of measuring tasks and required automation level the structure of automatic systems can vary a great deal. While large measuring systems can contain more than 50 instruments or devices, measurement set-ups containing as few as two or three devices are very common.

There are also considerable differences in the way in which measuring systems are realized. The assembly of the system can be undertaken by the user or by the manufacturer; the arrangement can remain unaltered for a long period of time, or can be required to be changed on a daily basis. In most cases the instruments are in the vicinity of the controller; however, in some instances it is unavoidable that the measurement or display is remote from the main system assembly.

There are great differences in the operational speed of individual measuring systems. Fast data-collecting systems containing special analogue-to-digital converters operate in the Mbyte/second range, while the operational speed of automatic measuring systems containing traditional instruments – such as voltmeters, frequency meters, etc – may be only a few kbytes/second.

In Chapter 1, while discussing the classification of automatic measuring systems, it was pointed out that there is a fundamental difference between 'externally-programmed' (e.g. punched-tape-controlled) and 'internally-programmed' (e.g. computer- or programmable-calculator-controlled) systems.

Let us examine, using a practical example, the significance of the two different control modes to the user. Let us assume that the task is the determination of the noise voltage of a VHF receiver as the function of the unmodulated RF input signal. The flow diagrams shown in Figure 3.1 illustrate the two different control methods.

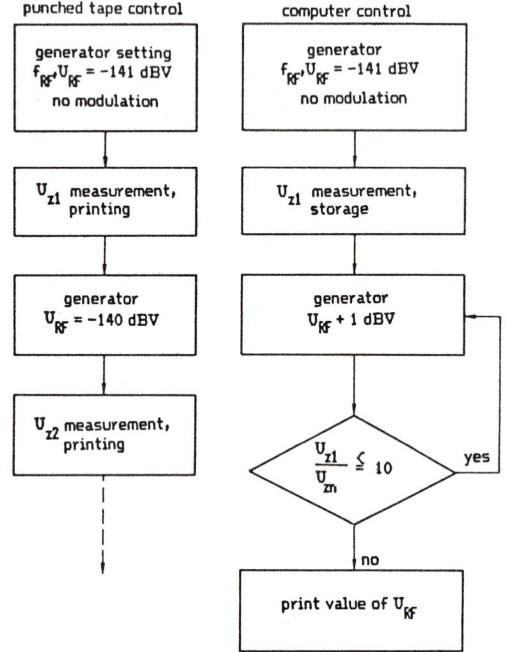

Figure 3.1 Flow diagram of punched tape and of computer control

The external programming is the simplest control method. It can be used only for measurement tasks not requiring comparisons or logical decisions. Using a punched-tape-controlled measuring system this measurement consists of one sequential series, as illustrated in the Figure. The output level of the signal generator connected to the receiver is incremented by 1 dB in each step and the noise voltage and the results are printed out. If we want to find out the RF level equivalent to 20 dB decrease in the noise voltage, then the evaluation of corresponding $U_{RF}-U_z$ values of a long data series is necessary.

In comparison the computer-controlled measuring system solves the problem in a much more advanced manner, using the least number of steps and printing out the final result. This is due to the fact that the computer is capable of comparing the stored result (U_{z1}) and the actual result (U_{zn}) and making a decision whether to continue the program, depending upon the result of the comparison.

As is apparent from the example, the characteristics of the controller, the speed of comparison and decision making, the data and program storage capacity and the simplicity of the interfaces connected to the instrument all greatly influence the productive capacity of the entire measuring system.

The computing industry – producing various controllers, mini- and microcomputers and calculators – is perhaps the most dynamic development area of electronics. One main feature is that the development of measuring and computing technology is becoming more and more interconnected.

The instrument designers utilize the advantages or programmed control in new-product development, while the special aspects of measuring technology application has begun to influence the development of minicomputers and calculators. The lower cost of computers – owing to the changes in manufacturing technology – has made measuring automation viable in practice. One must not forget, however, that the present technological level is a result of a lengthy development process.

3.1 Development of computer control

The history of modern computing technology began in 1946. It was then that the result of three years work came to fruition as the first digital computer was made on the premises of Pennsylvania University. The designers christened it 'ENIAC', which is the acronym of 'Electronic Numerical Integrator and Automatic Computer'. The weight of the apparatus exceeded 30 tons, it had 18 000 thermionic valves and consumed 130 kW of energy. The enormous energy demand posed a major problem for the designers of the ENIAC and other so-called 'first-generation' computers.

In the beginning of the 1960s a technological revolution took place in computer manufacturing. The first-generation valve-operated computers were replaced by the second-generation computers which contained transistors and diodes.

The closer involvement of measurement and computer technology began in 1963. In that year the first minicomputer appeared. Unusually small size and relatively low cost were some of the features of the Type PDP-5 computer, produced by the Digital Equipment Corporation.

The appearance of the minicomputer meant a major change in the use of computers in measuring technology. Up till then large capacity and very expensive computers were used in an off-line operational mode. This meant that the data collected by the measuring system were transferred to the computer at the end of the measurement series in a batch format on magnetic tape or punched card (Figure 3.2(a)). The results were available in the required format after the processing time, which depended upon the magnitude and priority of the task and on the other loads on the computer.

As opposed to this, the comparatively cheap minicomputers were used in an on-line operational mode from the very beginning (Figure 3.2(b)). In this operational mode the input/output unit of the computer is directly connected to the measuring

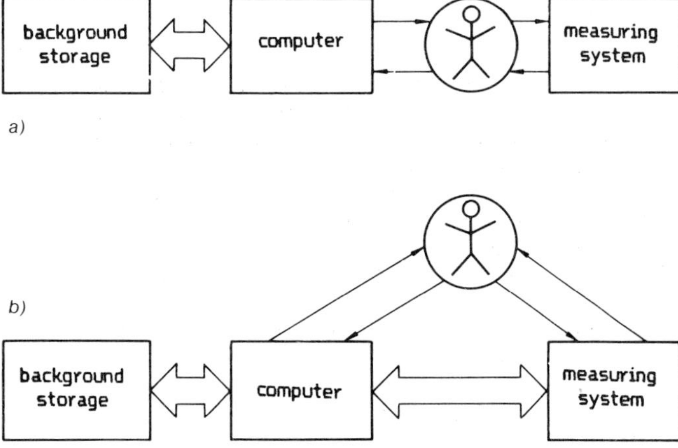

Figure 3.2 Computer data processing in the various operational modes. (a) Off-line, (b) on-line

system and the processing is virtually simultaneous. The role of the person controlling the measurement also changes, being removed from the data traffic and gaining the possibility to control the processing. The direct application of modern computing technology in the field of measurement automation was therefore made possible by the availability of these minicomputers and the development of automatic measuring systems accelerated to a great degree after this. The fast operational speed and high accuracy provided by the minicomputer control and the varied software possibilities became very attractive for the users.

Minicomputers then became progressively larger in capacity and smaller in size and steadily cheaper and two years after the appearance of the PDP-5, Digital Equipment Corporation introduced their Type PDP-8 minicomputer (Figure 3.3). Up to the date of first writing this book, the series PDP-8 was the largest-selling minicomputer family manufactured anywhere. Up to the end of 1976 more than 30 000 units of the PDP-8/A alone were sold.

At the end of the 1960s the semiconductor manufacture had a breakthrough to a new, so-far-unexplored field and the manufacture of semiconductor computers began. The first major success in the ferrite storage *v* semiconductor storage competition was achieved by the Type 1103, 1024-bit capacity, dynamic RAM (random access memory) produced by Intel.

In 1970 it was reported by MOSTEK, then a little later by Texas Instruments, that a calculator

Figure 3.3 Digital Equipment Type PDP-8 minicomputer

containing a single integrated circuit had been manufactured successfully.

The next stage was marked by the appearance of programmable calculators. These units were capable of executing complex operations by the pressing of a single key, using a program built into the hardware. Another advantage was that the calculators were operative immediately when switched on, while computers can be used only after the compiler or interpreter program has been loaded. However, the relative disadvantage of calculators is that their operation is comparatively slower and only a few have an interrupt system.

The efforts in calculator development lead to the microprocessors that later revolutionized computer technology. Intel produced a central unit in a single IC chip, known as the 'microprocessor', in 1971 during the development of the circuits of a programmable calculator. The prototype 4-bit 4004 microprocessor was followed by an 8-bit 8008 shortly, that was capable of handling alphanumeric data.

The development of semiconductor manufacturing techniques, especially the continuous improvement of MOS (metal-oxide semiconductor) manufacturing technology, fundamentally changed the structure of the central units of computers. Conditions for the manufacture of MOS improved markedly during the early 1960s. Although the principle of field-effect, or unipolar transistors, that is the basis of the MOS circuits, had been discovered in 1926, they could only be made about 30 years later due to the demand for surface purity and exact dimensional tolerances in their manufacture.

MOS circuits have some special characteristics that differentiate them from bipolar devices. They can be used in a wide range of supply voltage, and their power consumption is considerably smaller. Their advantages in manufacture are that they are self-isolating, have bilateral symmetry, can be used as either active or passive devices, and there is only a small percentage of rejects in their manufacture. An MOS transistor occupies only about the fifth of the space on the silicon base-plate required by a corresponding bipolar transistor device.

The major aspiration in the field of microcomputers using MOS technology is the increasing opportunity for component integration, leading to a decrease in the number of circuit elements. Figure 3.4 illustrates the development trend of 8-bit microprocessors marketed by one of the leading microcomputer companies, the Intel Corporation.

The 8008 family was made using the so-called first-generation P-channel MOS technology and the central unit was marketed in a simple 18-pin DIP (dual in-line package) chip form. Its 12.5 μsecond cycle time and its command set, consisting of only 48

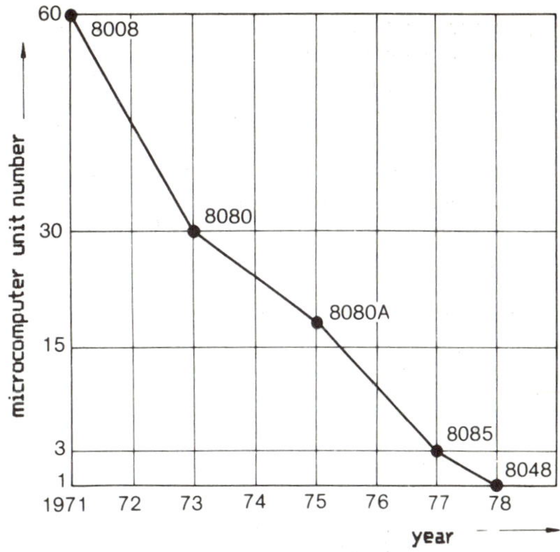

Figure 3.4 Development trend of the Intel 8-bit microcomputer families

commands, may seem humble now, but it was adequate in surprisingly many application areas. One complete microcomputer consisted of about 60 integrated circuit chips arranged on a large board.

The experience gained in manufacturing the N-channel, high component-density RAM unit and in developing the 40 output DIP chip led to the next stage. In 1973 Intel marketed their Type 8080 microprocessor, which was upwards software compatible with its predecessor, the Type 8008. This enabled the 8008 users to minimize much of the costs already invested in software development on the simpler system, while using a considerably improved new central unit (78 commands, 2 μsecond cycle time). This played a major role in that the Type 8080 virtually became an industrial standard, and for a long time more of these were marketed than the combined sales from all other different type microprocessors. Apart from the larger capacity, the number of discrete chips constituting the microcomputer dropped to half of its original value using this central unit. By integrating some functions this number was reduced even further, as shown in the case of Type 8080A and Type 8085, the latter being software compatible with the former.

A peculiar situation arose when, due to the accelerated drop in the costs of microprocessors, the manufacturing costs of microcomputers largely consisted of the costs of manufacturing the printed circuit boards. Further development in circuit integration and certain internal management alterations led to the manufacture of integrated single-chip microcomputers.

These devices are naturally of minimal structure; for instance the Intel 8048 contains a 1-kbyte storage capacity program memory and a 64-byte data memory integrated within the central unit.

MOS microcomputers became widespread very quickly, mainly due to their low cost. Unfortunately a common disadvantage of MOS circuits is that their operational speed is comparatively slow. The reason for this is that in MOS circuits during logic level changes, the charge and discharge of parasite capacities loading the inputs and outputs takes place via very high resistances.

In applications requiring very rapid operation the so-called bit-slice microprocessors, made using the bipolar Schottky TTL or ECL technology, gained advantage. In bit-sliced microcomputers the 8-, 16- or even 32-bit central units can be built up from 2- or 4-bit elements. This system organization not only allows fast speeds and an arbitrarily-extendable module-type structure but also provides a possibility of introducing another very advantageous method, microprogrammed control. Microprogramming means that the command set of a system is defined by microcommands stored in a fast-operating storage, the so-called 'microprogram memory'.

One advantageous feature of microprogrammed, bit-sliced microcomputers is that they can emulate MOS microcomputers. This means that the fast bipolar based microprogrammed computer can produce the entire command set of the MOS microcomputer that is normally provided with extensive software. The main advantages of the two different systems can thus be combined.

3.1.1 IEC interface units

The control units used in automatic measuring systems can be classified in four groups, although these are becoming less and less distinct:

- minicomputers,
- microcomputers,
- calculators, and
- special controllers.

The international acceptance of the IEC interface system induced the computer manufacturers to work with this standardized system. To be able to appreciate this, it is helpful to bear in mind that the various manufacturing firms in computer technology did not strive for standardization. The large minicomputer firms adhered to their own successful interface systems which did not have many characteristics in common.

For example, the I/O bus of the Data General Nova family contains six address lines, while in the bus system of the Digital Equipment Unibus (used in the PDP 11 series computers) there are 18 address lines. Similarly, the Interdata and Hewlett-Packard companies each use an entirely individual input/output unit in their minicomputers.

The acceptance of the IEC system is the consequence of the marked interest of these firms in measurement technology, representing a growing market. Appropriately they competed with each other to launch the IEC interface unit. The proposers of the interface system, Hewlett-Packard, had been the first in this field, closely followed by Digital Equipment, Computer Automation and even by the mainly data processing computer manufacturing IBM.

A similar phenomenon can be observed in the microcomputer field as well. IEC interface units, realized as LSI integrated circuits, have been manufactured for Intel 8080, 8085 and 8048, Motorola Type 6800 and other microprocessors.

As well as the computer manufacturers, lately some companies (mainly instrument manufacturers) have also been providing IEC control units. These special IEC system controllers are basically optimized minicomputers equipped with built-in display and floppy-disk unit for instrument control tasks. These types of control unit are made by the American Systron-Donner, Siemens and Philips, amongst others.

3.2 Minicomputers

There are two main groups of electronic computers. In analogue computers the information is represented by an analogue characteristic, a continuously-variable voltage value, while in digital computers two discrete voltage values (H and L) hold the data to be processed in binary form (1 and 0). Hybrid computers are also manufactured and these are combinations of analogue and digital computers.

In practice digital computers have gained the widest usage, so much that in everyday language the term 'computer' is used for digital computers almost exclusively.

As shown in Figure 3.5 the three main components of a computer system are the 'hardware', the 'software' and the 'firmware', the latter found in more recent models.

Hardware is the collective designation applied to the physical units of the computer system. The electronic, electrical, mechanical and magnetic components of the computer belong to this. As shown in the Figure, the main components of the hardware are the operation-controlling central unit, the storage containing the data and commands and the input/output unit providing the connection between the computer and the peripherals.

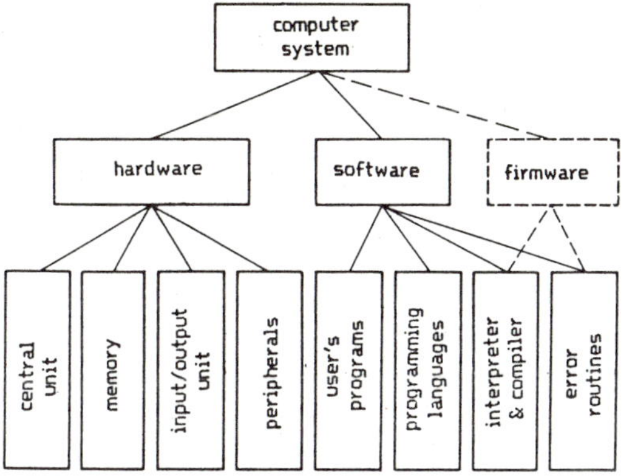

Figure 3.5 Computer system structure

The other main component of computer systems is the *software*, comprising the programs provided by the manufacturers and those developed by the user. Often the software is classified as 'system software' and 'utility' or 'application software'. The system software contains the operational system essential for running the computer, while the utility software consist of the user's programs.

In modern computers there is frequently a third component, the *firmware*, which comprises the programs permanently stored in semiconductor stores. In other words, the firmware is a software task carried out by the hardware. Usually the permanent program sections (e.g. loader programs) belong to this category.

One feature of the development of computing technology is that the proportion of hardware/software costs is changing all the time. This hardware/software ratio had been around 9:1 for the first-generation valve-operated computers, while at present the ratio is about 3:7.

3.2.1 The characteristics of minicomputers

There is no universally accepted definition for 'minicomputers'. Usually general purpose computers belonging to the minicomputer category have purchase prices not exceeding $20 000 in basic configuration.

The main characteristics of minicomputers include the length of the data word, the form of storage medium and its access speed, the structure and cycle time of the central control unit, and the means of communicating information in and out of the system. Each of these features must be considered jointly with each other to estimate the overall system response. Each item is considered in the remainder of this section.

3.2.1.1 Data format

One of the most important characteristics of a minicomputer is the word length, that is the number of bits that can be read from or written to the main storage during a single cycle. It is generally true that longer word length ensures greater accuracy and more efficient operation, although causing higher costs. Most minicomputers are of 16 bits. This word length represents a very advantageous compromise between capacity and cost.

The instruction length is a similarly important feature and is either the same as the word length or a multiple of it. In the two-word instructions the first word is generally used to identify the operation to be carried out while the second contains the address of the operand. The use of instructions containing two words increases the directly addressable storage capacity, resulting in simplified programming.

3.2.1.2 Main storage

In minicomputers the main storage used for saving the data and the instructions is either magnetic core (ferrite ring) or semiconductor type. The non-volatile ferrite storage devices are being steadily replaced by the small and cheaper semiconductor storage. The latter have an important disadvantage, however, in that the information stored in them is lost when the power supply is switched off or interrupted even for a short time, as distinct from the ferrite devices which permanently retain their data. The various types of semiconductor storage devices and their characteristics will be discussed in the section introducing microcomputers.

Cycle time is one of the most important characteristics of computer main storage; it represents the shortest time elapsing between the start of two consecutive storage access cycles.

Access time is another important characteristic, giving the time delay between the allocation of a particular storage address and access of the required information.

Capacity of the main storage is not a fixed characteristic of a particular computer. Manufacturers usually give the minimum and the maximum storage capacities and within these, the user has to decide how large a capacity is required by his biggest program, including subroutines and data. In up-to-date computers the storage capacity can be readily increased by the addition of plug-in storage modules.

Parity checking is a permanent feature of some minicomputers, while in others it is an optional extra, depending on the user's demand. During parity checking a 0 or 1 value bit is generated and assigned to individual storage sections and with the aid of these, storage errors are recognized during read-out.

3.2.1.3 *Central unit*

The structure of the CPU (central processing unit) of individual minicomputers varies a great deal. The majority of them carry out the arithmetic operation in parallel with the storage word, which is usually 16 bits long.

Directly addressable memory is, from the user's point of view, one of the most important characteristics. A typical 16-bit minicomputer instruction consists of three parts which are: the operation code field, the addressing mode field and the address itself. If 6 bits represent the operation code (permitting 64 different operations) and 2 bits are used for signalling the addressing mode, then 8 bits remain for writing the actual address. With this 256 different storage words can be directly addressed.

The remainder of the computer's main storage can be accessed using other addressing modes (e.g. indirect, indexed, relative, etc). Of the various addressing modes the most important is *indirect addressing*. With this process the address field of the instruction assigns a storage word that does not store the operand, but another address. This second address is either the address of the operand or it is another indirect address. The latter case represents a multi-level indirect addressing.

The manner in which the control mode is realized is an important characteristic of the central unit of a minicomputer. The structure and the operation of a traditional cable connected logic structure central unit cannot be altered. In contrast, the operation of a computer can be modified when the contents of the storage of microprogrammed central units are changed.

In comparisons of computer operational speeds the arithmetic operational time can be misleading. Despite this, a *fixed decimal point addition time* is useful, so as to give a guide in judging the arithmetic speed of the computer.

The multiplication and division process method is an important characteristic of the central unit. It is considerably faster when these operations are executed by the hardware, rather than by the software using programmed subroutines.

The *floating decimal point arithmetic unit* is similar to this. These units, handling exponential numbers are essential in measurement technology applications, as they free the user from the task of changing the decimal value of the numbers.

3.2.1.4 *Input/output control*

A basic concept concerning the input/output operation of minicomputers is that of *Direct Memory Access* (DMA). If a minicomputer does not possess a DMA channel then the transfer of each word between the peripherals and the main storage takes place using program control via the registers of the central unit.

In theory the maximum data transfer (I/O) rate is determined by the cycle time of the main storage in the case of minicomputers equipped with a DMA channel. In practice, however, the capacity of the main storage and the lack of simultaneous input/output operations generally limit this.

The possibility of *program interrupt* is an essential condition for the real-time operational use of computers. The interrupts can belong to either of two categories, internal interrupts or external interrupts. The former can be caused, for instance, by a parity error, illegal instruction or power supply failure. The latter are usually in response to a service request by the peripherals. The number of external interrupt levels provides information about the efficiency of the computer's interrupt system.

3.2.2 Peripherals

The money spent on the central unit is only a small part of the hardware costs of the computer. The peripheral units are considerably more expensive and are generally less reliable in operation. For these reasons particular attention should be paid to peripherals when selecting the units of a computer system.

The peripherals consist of various background storage devices and of data input/output units. In modern minicomputer systems magnetic disk units are used for the storage of large amounts of data and for programs requiring large storage capacities. There are two variants of the traditional hard disk: *fixed disk* and *interchangeable disk* stores. The advantage of the interchangeable disk is that, due to the interchangeable disk packages, it represents unlimited storage capacity, in theory. However, the advantage of the fixed disk store is that it permits considerably faster access to data. Data is stored in different sections and a separate write/read head is allocated for each sector; therefore, the only delay during input data writing or output data reading consists of the time required for the selected part to reach the head. In comparison the interchangeable disk uses a single head which has to be moved to the appropriate section. The fixed disk stores are more expensive, and therefore their use is economical only when fast access to large amounts of data is necessary. The capacity of hard disk stores in between 5 and 500/Mbytes and the average access time is typically 0.01 to 0.1 second.

Floppy disks, or diskettes, are also frequently used as background stores for modern minicomputers. Whilst these units are slower and have smaller capacity than the hard disks, they are considerably cheaper and their maintenance is easier. The capacity of floppy disks is between 200 kbytes and 10 Mbytes and the average access time is 0.5 to 1 second.

At present the cheapest background stores are magnetic tape cassettes. Although recently there has been a major development in their technical characteristics, they still lag behind magnetic disk units in capacity and operational speed, with capacities between 60 and 500 kbytes and access times of 10 to 300 seconds.

In measuring technology applications where the input/output peripherals provide an interactive operational mode, the console keyboard and alphanumeric display units play important roles. These peripherals are connected to the computer via a serial interface (V24/RS-232-C). The data transfer rate of the console keyboards is fairly slow at 110 baud (bit/second); however, the display units do not contain mechanical parts and are therefore much faster, permitting data transfer rates in excess of 9000 baud.

The output device of a minicomputer is usually some form of printer. These are either matrix printers or line printers. Matrix printers which print a single character at a time, are used when there is no need for the printout of long lists and where the printing speed is not critical. These units typically print from 30 to 600 characters per second. Line printers are more expensive devices, printing 80 to 160 characters at once. The operational speed of these devices is 100 to 2000 lines per minute.

The data input/output units discussed so far are for handling alphanumeric data. In measuring technology, however, *graphic peripherals* – suitable for displaying analogue information – are also essential. The most important of these are the graphic display units using cathode-ray tubes and these can be classified in three groups according to their operating principles. This determines, to a great extent, the characteristics and the operational modes of the units.

In the *storage-tube display* devices the graphic information is displayed on a bi-stable storage cathode-ray tube. The images of these displays are of good quality and flicker-free, and service software is comparatively simple. Their disadvantage, however, is that only the entire image displayed on the screen can be erased.

In the *raster-scan displays* the image is drawn by scanning the entire screen and it is updated continuously using a frequency of 30 to 50 Hz. The details of the drawn images can be modified in the interactive operational mode. This type of display unit is considerably more expensive than the storage tube variant.

The *direct-beam display* forms the third group of the graphic displays. Its individual image segments can be displayed using various light intensities and the image can be overwritten very quickly. In this type of display the complexity of the image drawn depends on the memory available and the complexity of the special software controlling the display.

The *digital drawing devices*, or X–Y plotters, constitute the other group of graphic displays. In these units the drawing of the images is achieved by electromechanical or electrostatic methods using digital data expressed in X or Y coordinate values. In modern plotters microprocessor control simplifies and reduces the external control tasks. This is a general trend and a reduction of loading on the control computer is attempted in this way whenever a 'slow' peripheral is designed.

The ever-decreasing physical size of the central units of computers and of the various peripherals creates new possibilities for designers. More and more minicomputer manufacturers are marketing desk computers that contain the most essential peripherals and the interpreter programs in ROM. The Compucorp Type 625 desk computer, shown in Figure 3.6, is an early but typical example of this. It contains an alphanumeric keyboard and printer, as well as a 1024-character display and a 600-kbyte capacity floppy disk integrated with the central unit.

3.2.3 Software

The software costs of minicomputer systems exceeds the hardware costs several times over. The software controls the operation of the computer; therefore, it can be said that the efficiency of the software and the supply of suitable programs are the determining factors in the capacity and universal applicability of the computer.

This must be borne in mind when a particular type of computer is selected. The operational speed of the central unit or the size of the directly addressable memory provide some information about the state of development of the hardware, but these become far less significant compared to the importance of the availability of suitable software.

Unfortunately, software systems can be compared only with difficulty, because the only people who can reliably form an opinion of the software system of a given computer are those who have used them, i.e. have written programs in the available languages and have run them on the computer. It is for this reason that computer magazines publish users' ratings to facilitate the task of computer buyers.

However, even these users' opinions do not provide a clear guideline since the various computer programs are optimized for particular purposes. For example, an interpreter program optimized for

Figure 3.6 Compucorp Type 625 desk computer

storage capacity may be regarded by many users as excellent, but may be totally unsuitable for ease of handling a small amount of data and providing the shortest possible running time.

An essential part of computer operation is the so-called 'operating system'; that is, the programs preparing and executing the user's programs. This operating system is a series of inter-related, mutually-complementary programs, written in a hierarchical manner. These programs belong to two categories:

- translator programs, and
- control programs.

The *translator programs* ensure the interpretation and testing of programs written in symbolic languages. Programming of the minicomputers can be undertaken on various program language levels. Digital computers are binary systems and their central units are therefore programmed in binary code to execute various operations. This code, containing the 0 and 1 values, is the machine code and programs written in this language are *machine code programs*. However, programming in machine code is so time-consuming and difficult to follow that it is rarely used in practice.

Translator programs facilitate programming for the user and a summary of these is shown in Figure 3.7.

The assembler translates a program written in symbolic code to the binary language of the computer. The mnemonic code of the assembler is easier to handle than the machine code; for example, the CLA (clear the accumulator) assembler language instruction of a PDP-8 computer is easier to remember and less likely to be confusing

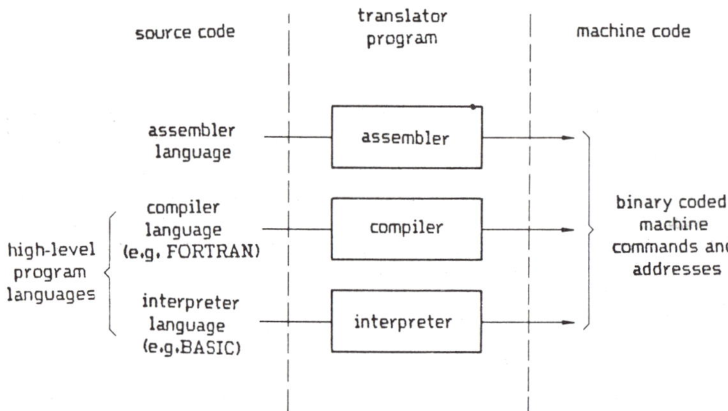

Figure 3.7 Various translator programs

than the corresponding 111110000000 machine-code instruction. Each assembler program instruction is equivalent to a machine-code instruction. The assembler is dependent upon the central unit, a different assembler language being required for each different computer.

Although assembler-language programming is considerably simpler than machine-code programming, a thorough knowledge of the operation of the central unit and extensive programming experience is required. Accordingly, high-level languages that make programming independent from the manufacturer and from the computer type were developed.

A common feature of high-level programming languages is that the programmer does not have to specify every elementary operation. The operations given in the *source code* are broken up into elementary machine code instructions which can be interpreted by the central unit in the *object code*.

Several machine-code instructions belong to a single instruction of the high-level languages. The use of these languages eases the task of the programmer considerably but, as an inevitable consequence, the program in question occupies considerably more storage than would be necessary for an equivalent assembler-language program. Despite this the program development time is reduced to such an extent that in practice programming of minicomputers is done almost exclusively in one of the high-level languages.

In measuring technology FORTRAN and BASIC programming languages are those most frequently used.

FORTRAN (FORmula TRANslator) is a programming language developed by IBM, several variants of which are used. FORTRAN IV, standardized by ANSI, is the most common of these. In measuring technology FORTRAN is used for research tasks requiring complex calculations, for which it is particularly suitable.

BASIC (Beginners All-purpose Symbolic Instruction Code) is a programming language developed by Dartmouth College in America, originally for teaching purposes. It has gained very wide acceptance and extended variants of the original BASIC are frequently used. All variants are interactive and interpretative, that is the instructions are interpreted, tested and immediately executed by the interpreter program line-by-line.

The other part of the minicomputer's operational system consist of the *control programs*. The task of these is to control the operation of the computer's central unit and peripherals in an optimal manner. In addition these programs permit the interaction between the computer and the operator. The most important control programs are the monitor, the loader, the I/O control system, the interrupt routine and the diagnostic program. With the exception of a resident part that is always in the main storage, the control and process programs are located in background storage.

3.2.4 Input/output units

The input/output subsystem of a computer should execute the following operations:

(a) Issuing commands to the peripherals.
(b) Transferring state signals to the central unit.
(c) Two-way data transfer.

The external peripheral devices of computers used in measuring techniques are collectively called input/output devices and there are two main categories of them:

- computer peripherals, and
- analogue measuring devices.

The programs and the units either providing data input or visualizing the results of computer processing (i.e. teletypes, displays, line printers, various background storages) belong to the category of the computer's data input and output devices.

The analogue-to-digital converters (ADC) and the digital-to-analogue converters (DAC) and those programmable instruments that supply measurement data in digital format, controllable by digital signals, belong to the analogue measuring devices.

For the synchronized connection between the central unit and the I/O devices an interface and a unit controlling the connection is necessary.

The basic difficulty of the I/O operations between the computer and the peripheral devices is that the computer's central unit – controlled by an internal clock signal – is much faster than the peripherals. For example, a teletype handles 110 bits per second, while the computer's main storage has a data transfer speed exceeding several Mbytes/second. The role of the timing strategy of I/O operations is to synchronize the operation of the computer's central unit and the I/O units.

3.2.4.1 I/O operations

There are two I/O connection methods according to the organization of the data transfer: programmed and autonomous. Each method has advantages and disadvantages; for example, the fastest data transfer speed achieved varies a great deal. In the following section the characteristics of the two methods will be discussed in detail.

3.2.4.2 Programmed I/O connection

The programmed I/O is the simplest transfer method. In essence, the central unit issues a separate instruction for the transfer of every data item in the data traffic involving the I/O devices connected to the central unit via data, address and control signal lines. The programmed I/O connection can be executed using either parallel poll or program interrupt.

3.2.4.3 Parallel poll

In the case of parallel poll, the data transfer is handled by the current program (Figure 3.8). The

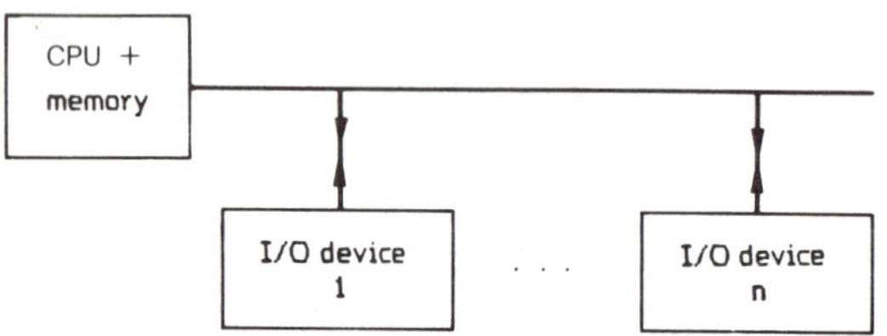

Figure 3.8 Block diagram of the programmed I/O connection

currently-executed program asks the I/O devices periodically whether or not they request service (Figure 3.9). If the answer is 'yes', then the service of the device is executed with the aid of the service routine, generally ensuring the transfer of a data word or data block. The central unit first examines

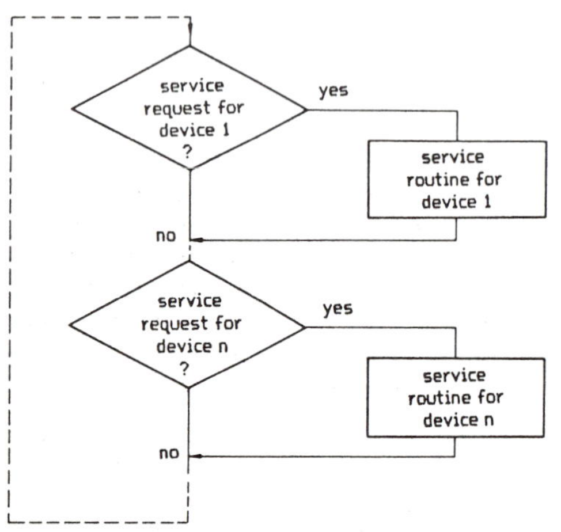

Figure 3.9 Flow diagram of the programmed I/O connection

whether the I/O device is capable of the data transfer or not and then transfers the first datum with an I/O instruction. The I/O devices are usually much slower than the central unit and therefore the issue of the next I/O instruction must be delayed. This delay can be created by using a loop incorporated in the program, but this paralyses the operation of the central unit completely. It is a better solution if the central unit executes another task during the waiting time, but periodically asks the I/O device whether it is now ready to receive new data. If this is executed by the devices on their allocated signal lines, then the identification of the device occurs at the same time and therefore the central unit can issue the next datum. Using the parallel poll an input device can also signal that it wants to transfer data to the central unit.

The periodic poll must be programmed in the central unit and it is therefore occupied with I/O operations most of the time, due to the corresponding software overhead. In practice this is not a tolerable situation and a better solution is for the I/O device to signal the data transfer ready state with a program interrupt.

3.2.4.4 Program interrupt

The I/O device initiates the program interrupt with a signal sent on the service request or interrupt line (Figure 3.10). If service request is not forbidden by

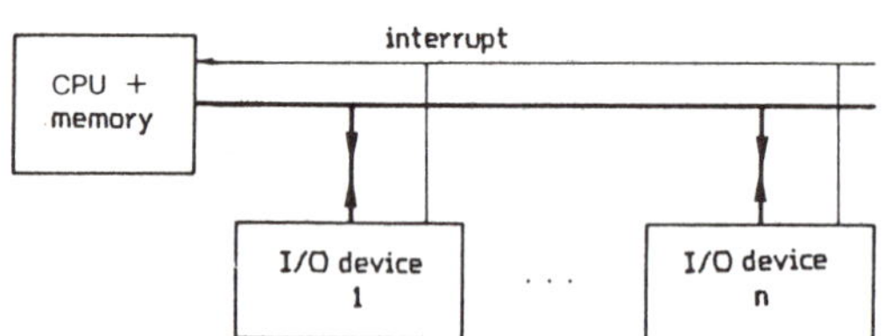

Figure 3.10 Block diagram of the interrupt-controlled I/O connection

the central unit's program and the central unit is in a state capable of saving the program being currently run, then it accepts the service request and starts the interrupt program. First it stores the states of the central unit and the program, then identifies the device requesting service and asks for the state word of the device to establish the reason for the interrupt. The central unit makes a decision about the execution of the device servicing program according to the state word. If an interrupt system assigns the start address of the service program the execution of the service program then becomes faster. This is called 'vectored interrupt'.

More than one device can request service simultaneously and a priority order must therefore be established to ensure appropriate operation. The priority order can be established by either software or hardware methods.

In the case of *software priority*, when an interrupt request is received, an interrupt handler routine polls the devices one-by-one. The poll – and if necessary, the service – of the highest priority device is executed first.

In the case of *hardware priority* the interrupt control circuit of the central unit sends a response signal to the I/O devices. This signal travels through all devices one-by-one and the priority is determined by the wiring order of the signal line.

The priority order has an important role in handling multiple interrupts. As shown in Figure 3.11 the individual interrupts are intertwined; in this case the return to the basic program is generally executed in the same order as the departures.

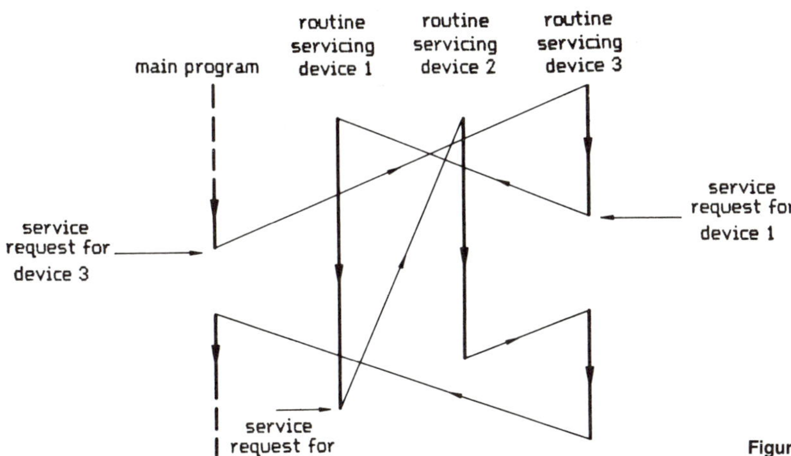

Figure 3.11 Handling of multiple interrupts according to priority

Although the interrupt serves the purpose of executing the I/O operations, all computers use external interrupts as well. These stop the currently running program when a special state occurs (for example, data transfer error, false instruction or the danger of power supply failure).

The advantage of the interrupt I/O data transfer lies in fast service handling, which is essential in real-time systems. One of the most important characteristics of a computer is the response time, i.e. the time that elapses between the reception of a service interrupt and the start of the service.

One disadvantage of the interrupt I/O is that the execution of the interrupt handling routines are not synchronous with the main program. This makes estimation of the memory sector capacities – and ultimately the programming – more difficult. A further disadvantage is created by the need for special hardware when hardware priority must be used.

3.2.4.5 Autonomous (program independent) I/O connection

In the case of the programmed connection the central unit has to execute several instructions for each data transfer and this uses a large portion of the central unit's time when data blocks are being transferred.

Using the autonomous connection the program assigns the storage address that is to be used as a starting point for the data transfer. Then the transfer continues independently from the program, in an autonomous manner. The transmission is terminated either by the transfer of a certain number of bytes, given in the program, or by the recognition of a last character. Hence, in the central unit the program of another task can be run during autonomous data transfer. Should the autonomous connection control and the central unit address the memory simultaneously, it is always the connection control that is granted priority and the central unit is left waiting for one memory cycle. This is called 'cycle stealing'.

With programmed connection all data must pass through either the central unit's accumulator, or through a register assigned for that purpose, while the autonomous connection is usually directly connected with the memory, avoiding the central unit.

3.2.4.6 Direct memory access (DMA)

The DMA controller was developed specifically for achieving direct memory connection. The aim of the exercise is to achieve autonomous transfer in the shortest possible time using the least number of program steps (Figure 3.12).

The data transfer begins with the programming of the DMA controller. The central unit assigns the start address of the memory and the number of words (bytes) to be transferred. These data are stored by the DMA controller. Then the central unit checks the state of the I/O device. If the device is ready for the transfer it requests service, using an interrupt given directly to the DMA controller. Then the DMA controller suspends the operation of the central unit, taking over the control of the memory addressing and of the entering (or reading) of data. Every transfer increases the address by 1 and decreases the value of the byte counter by 1. The transfer is continued until the byte counter becomes 0. Using this method the data transfer speed can be very fast, approaching the entry speed of the main memory.

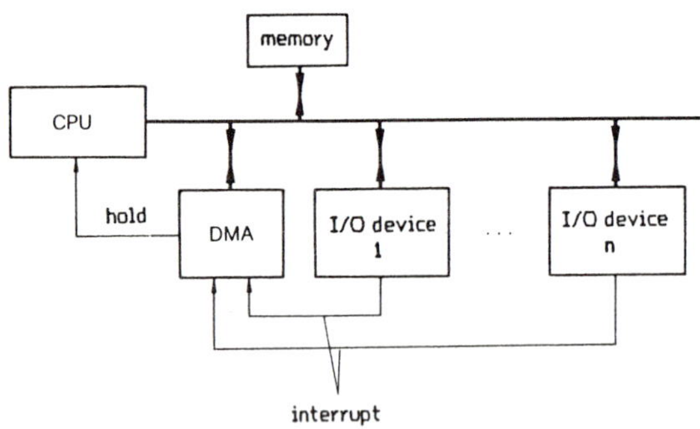

Figure 3.12 Block diagram of direct memory access (DMA)

3.2.4.7 The interface systems of minicomputers

Great variety has always been the hallmark of minicomputer parallel interface systems. To illustrate this, the parallel interface system of the two largest series of minicomputer families manufactured are discussed in the following section.

The parallel interface system of the Nova minicomputers, made by Data General and the unit providing interfacing to it is shown in Figure 3.13. This parallel interface contains a two-way data bus consisting of 16 data lines, six device selector lines, four test lines, 20 control lines and one line for setting up the start state. The interface operates synchronously, i.e. the data are strictly timed for both input and output.

In contrast to this, the parallel interface of the PDP-11 minicomputers of Digital Equipment Corporation operates asynchronously and its signal line system fundamentally differs from the Nova interface (Figure 3.14). The PDP-11 interface contains 16

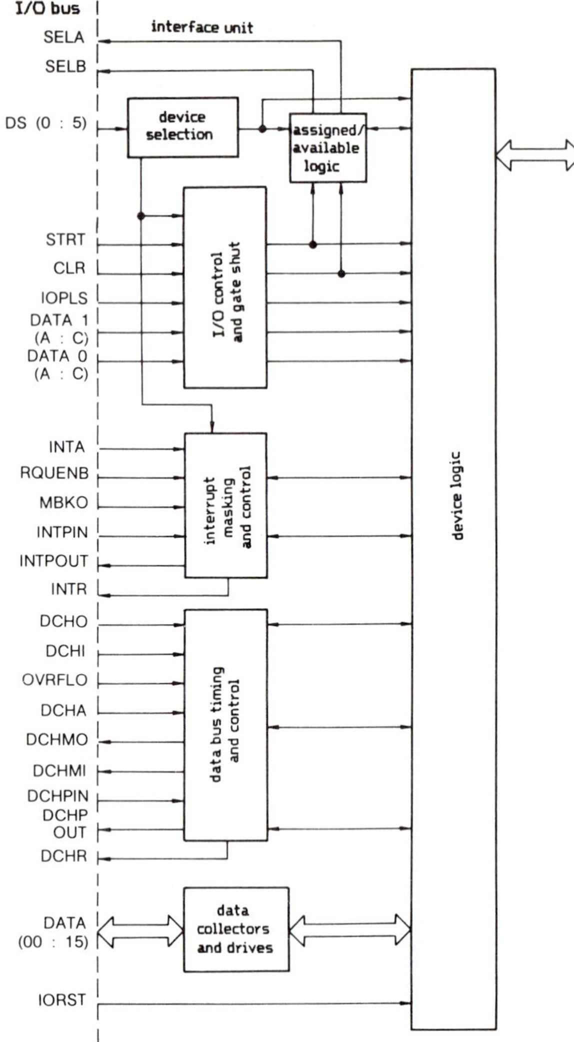

Figure 3.13 Parallel interface system of the NOVA minicomputers made by Data General

data lines, 18 address lines, 15 interrupt lines, six general control lines and one line for setting up the interrupt state.

These interfaces are but examples. Similarly, the standard interfaces of Computer Automation, Interdata, Hewlett-Packard and other minicomputer manufacturers are each fundamentally different.

Another interface system that is widely used in personal computer systems is a parallel interface containing 100 signal lines, developed in 1976 by the American MITS and IMSAI companies. A serious drawback of this so-called S 100 interface was that it was not defined accurately enough and therefore

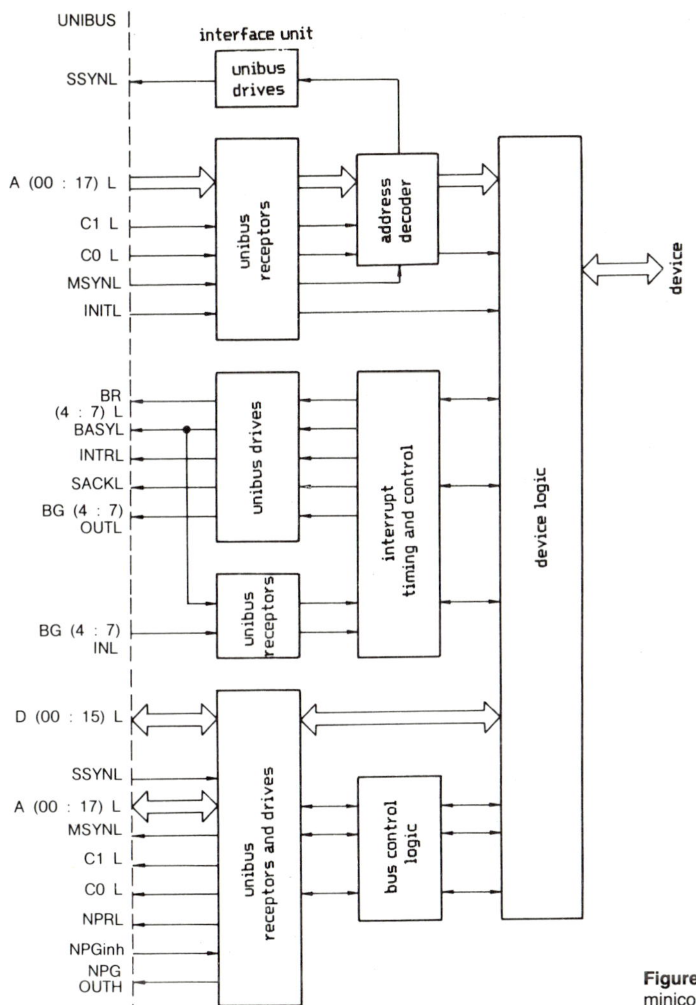

Figure 3.14 Interface system of the Type PDP-11 minicomputers made by Digital Equipment

various manufacturers did not use certain signal lines in the same way. This situation has now been remedied by a publication under the auspices of the IEEE.

The interfacing of slow operating I/O devices deserves a special mention. The majority of these possess a standard bit-serial interface and hence almost all minicomputers are equipped with an interface unit that connects the bit-parallel minicomputer interface to the bit-serial I/O device interface.

Of the bit-serial interfaces the 20 mA current loop was first used and later the EIA RS-232-C standard serial interface, which for our purposes is virtually identical with the international standard CCITT V24. In the RS-232-C interface system a character is constructed from 1 START bit, 7 data bits, 1 parity bit and one or two STOP bits (Figure 3.15). The value of the START bit is always 0, while the value of the two STOP bits is always 1. The 7 data bits contain the information, usually in ASCII (ISO 7-bit) code.

The RS-232-C interface operates in asynchronous mode, while the central unit of the computer operates synchronously. Hence, the serial asynchronous signal series are converted to parallel format in the central unit. The conversion is executed by a counter circuit. The counter starts a clock signal regenerating circuit in the middle of the 9.09 mseconds duration START impulse, then the sampling necessary for the conversion of the serial data occurs in the middle of the impulses incorporating the characters. The fastest data transfer speed achievable with the RS-232-C interface is 20 kbits/second. These and other interfaces are discussed more fully in a later chapter.

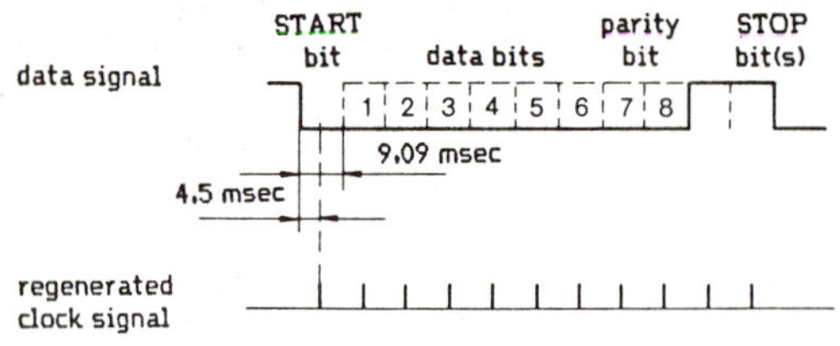

Figure 3.15 Serial data transfer in the RS-232-C system

3.2.4.8 *Interfacing of minicomputers and the IEC system*

In Figure 3.16(a) a minicomputer and three I/O devices connected to it are shown. All three devices are each connected by an individual interface card to the parallel computer interface driver. Similarly, three different handler/driver programs are required for the control of the devices. The disadvantages of this method can be seen clearly, if we consider that the design and development of an interface for a single device – if undertaken by the user – can take several months and the total costs can approach the price of the minicomputer.

This problem is especially noticeable in measuring technology applications, because the users are not well versed in computer interface design. Furthermore, the instruments used as I/O devices are comparatively cheap, making the high cost of the interface units even more disproportionate and prohibitive.

For these reasons the existence of the standard IEC interface is significant, allowing the connection of 14 I/O devices to the computer (Figure 3.16(b)). Almost all minicomputer manufacturers interested in measuring technology applications produce the IEC interface units and the corresponding software routines as standard options. The hardware part of the interface unit is usually placed on a printed circuit card which can be plugged into the computer. Figure 3.17 shows a Hewlett-Packard minicomputer and its interface unit.

In the following section the directives governing the structure of a minicomputer/IEC interface will be examined by using a practical example.

3.2.4.9 *The PDP-11/IEC interface*

The PDP-11 computer can be connected to the IEC bus via an IEC11-A type interface unit (Figure 3.18). The interface unit – constructed from medium-scale and small-scale integrated circuits (MSI, SSI) – can be connected to the UNIBUS via the standard I/O connector (CS1) of the PDP-11 family. The IEC bus interfacing uses a standard (CS2) connector (Amphenol 17-20250), while the card's third connector (CS3) provides interfacing for the optional DMA unit.

The data transfer between the interface unit and the computer is executed by program control, using program interrupts. Selection of the interrupt vector value and the assignment of the interface's UNIBUS address are set by switches located on the interface card. The hardware priority number of the interrupt and the IEC talker and listener addresses are set by additional selector switches.

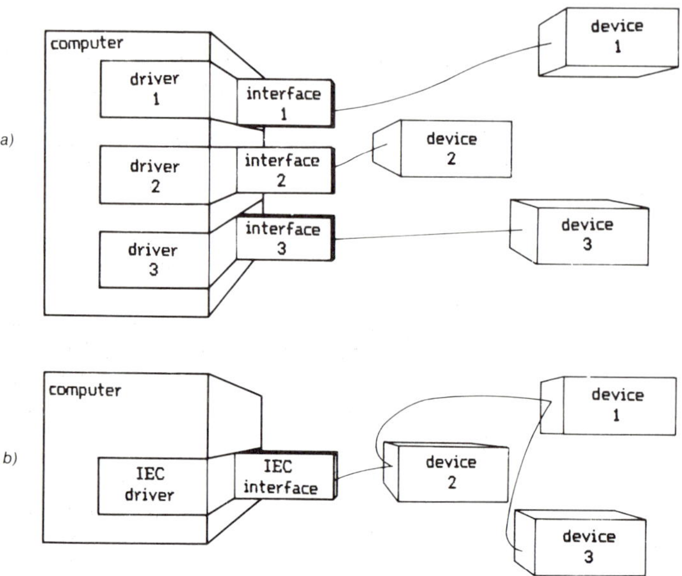

Figure 3.16 Interfacing of minicomputers to measuring systems. Various realizations: (a) Parallel interface, (b) IEC interface

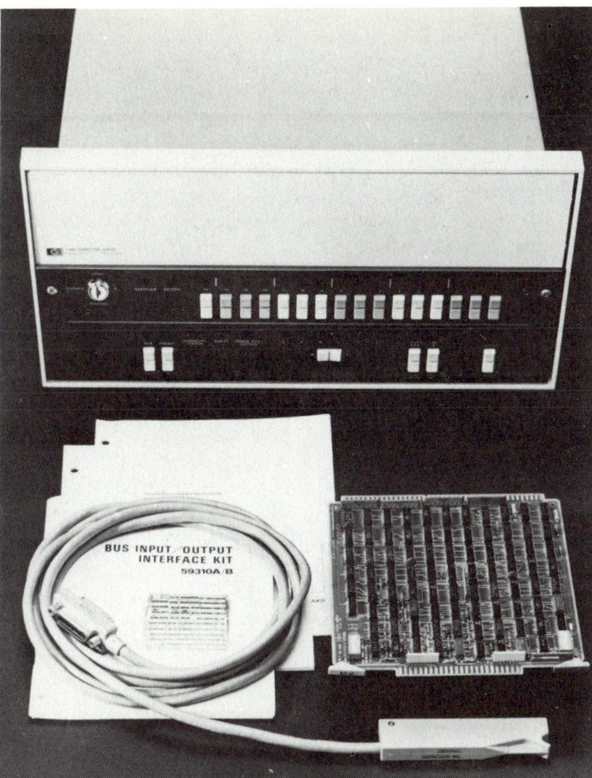

Figure 3.17 The Type 21 MX Hewlett-Packard minicomputer and its interface kit

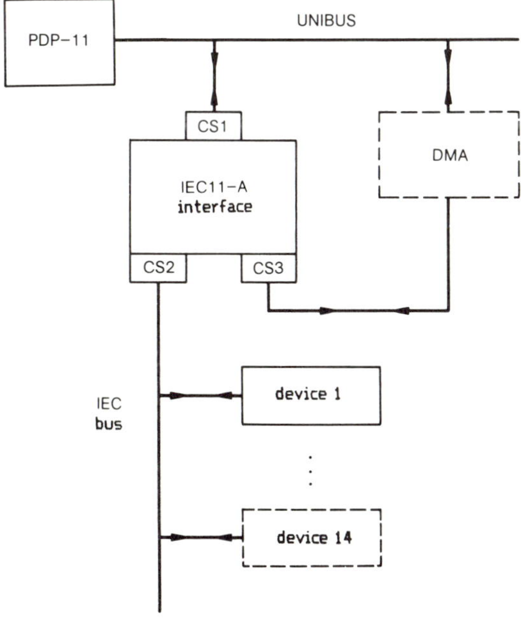

Figure 3.18 IEC interfacing of the Type PDP-11 Digital Equipment minicomputers

The greatly simplified block diagram of the IEC11-A interface is shown in Figure 3.19. The interfacing to the computer is essentially achieved by four registers that are filled by the program and can be interrogated. These are the following:

- The control and the interrupt register (CIR).
- The state and the message register (SMR).
- The input/output register (IOR).
- The vector switch register (VSR).

The CIR register mainly serves the purpose of the control of program interrupts. Accordingly, one of its control signals is the Service Request (SRQ) message of the IEC system. It also stores some of the special IEC control signals; one of the most important of these is the lon (listener on) signal.

The SMR register is the heart of the entire interface. This unit stores the local messages and is also used for assigning some interface states.

The IOR register consists of two symmetrical parts, one providing data output, the other data input. If the interface is in the Controller Active State (CACS), or in the Talker Active State (TACS) then the databyte or the instruction is transferred by the program from the computer to the output buffer and from there to the IEC bus. If the interface is in the Listener Active State (LACS), the databyte arriving from the bus is transferred by the program to the input buffer and from there to the computer.

The VSR register has an important role in handling the interrupts. This unit contains the address, set by the selector switches, to which the program jumps after an interrupt.

The interrupt start logic is also an important part of the interface functions that make up the signal controlling the interrupt start from the various IEC interface messages (EOI, SRQ, etc) and from local control signals (e.g. illegal message, state change, etc).

Figure 3.20 shows the contents of the registers used for the interfacing. These registers contain the values of the local messages and the states belonging to them. Any local message can get to logic 1 state only if the corresponding state is active. For example, the tca is effective only if the CSBS state is active.

The contents of the registers are closely interrelated. For example, the change of the interface states present in the SMR register causes the appearance of the STATE CHGE (state change) message in the CIR register.

The contents of the rest of registers are similarly interrelated. During the issue of commands the output data residing in the IOR register are transferred to the IEC bus during the active state of the CACS (SMR register). When new data are

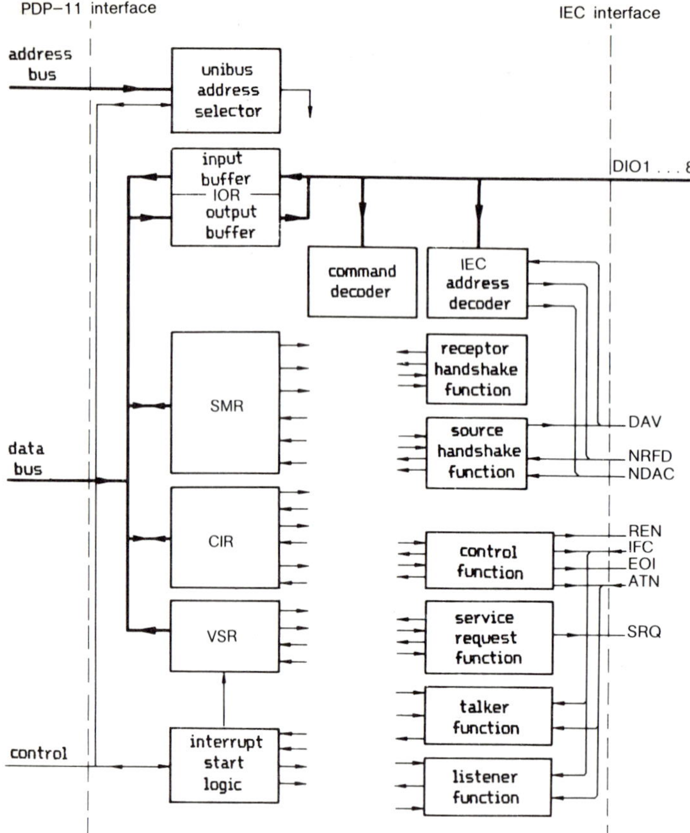

Figure 3.19 Simplified block diagram of the IEC-11A interface

input, the DATA ACC active state in the CIR register, signals that the command was accepted and the new byte can be put into the output buffer.

The IEC system compatible devices (including the controllers) use only part of the possible interface functions. The computer/IEC interfacing must always contain the control function, so as to be able to send messages to the devices connected to the IEC system and to react to the various messages sent by the interfaces devices (service request, etc). It is also necessary to transfer data in both directions between the computer and the IEC system and therefore the *talker* and *listener* functions, as well as the *acceptor* and *source handshake* functions must also be used.

Table 3.1 lists the functions and their variants used in the PDP-11/IEC interface. It is apparent from the Table that although this interface does not use every function, it can respond via the C function to the messages transmitted by any realized function of the system devices. For example, the computer can assign the *local* or *remote* state to a device via the C function, therefore the computer need not possess the *local/remote control* function. The same applies to the *parallel poll*, and the *device trigger* and *device clear* functions.

Table 3.1 The functions of the IEC Type 11-A interface

Symbol	*Possible functions*	*Realized variants*
SH	Source handshake	SH1
AH	Acceptor handshake	AH1
T	Talker	T2
L	Listener	L1
SR	Service request	SR1
RL	Remote/local control	–
PP	parallel poll	–
DC	Device clear	–
DT	Device start	–
C	Controller	C1, C2, C3, C4, C5

15	14	13	12	11	10	9	8	7	6	5	4	3	2	1	0	
DATA ACC	SRQ	EOI	ILL MSGE	NO LAC	NDAC LINE	NRFD LINE	STATE CHGE	INTER-RUPT	INT ENABLE	MASTER CLEAR	rsv	BLOCK DAC	LAST BYTE	lon	rsc	CIR
SPAS	SRAS	SIAS	LACS	TACS	CPPS	CSBS	CACS	sre	sic	lun	ltn	rpp	gts	tca	tcs	SMR
IN 8	IN 7	IN 6	IN 5	IN 4	IN 3	IN 2	IN 1	OUT 8	OUT 7	OUT 6	OUT 5	OUT 4	OUT 3	OUT 2	OUT 1	IOR
				VECT 11	VECT 10	VECT 9	VECT 8	VECT 7	VECT 6	VECT 5	VECT 4	VECT 3	VECT 2			VSR

Figure 3.20 Register block of the IEC-11A interface

3.2.4.10 Operational speed of the IEC controllers

The data traffic of the IEC system controller in most cases is considerably greater than that of the other devices. The tasks of the controller are the addressing and programming of all of the devices, the starting and stopping of the measurements, logging data, etc. Therefore, the operational speed of the control unit can significantly influence the speed of the whole system, especially if it consists of many devices and if the measuring task requires frequent controller interactions.

If a user wishes to control the automatic measuring system with a computer, then he presumably wishes to utilize its fast speed, otherwise a simpler and cheaper solution would have been sought, e.g. using a programmable calculator for the control unit. Hence, two of the most important aspects of the minicomputer are the speed of its operation and the degree of utilization of the central unit when used with an IEC interface. An unambiguous numeric answer cannot be given on this problem, as the operational speed is influenced by many factors. These factors and their effects will be examined in the following example, the RTE (real-time executive) operation system controlled Hewlett-Packard minicomputer.

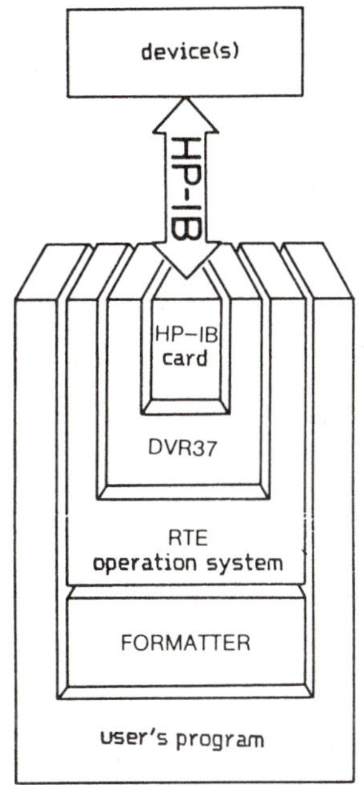

Figure 3.21 Software elements of the minicomputer/IEC interface

Figure 3.21 shows the software elements of the HP-IB interface. The connecting link between the computer and the devices is the HP-IB driver (DVR37) software routine. This transmits the commands generated at the operating system's level to the interface card and, conversely, the data arriving from the interface to the user's program.

As well as the DVR37 routine – which can be called upon via the FORTRAN, BASIC and assembler programming languages – there is a FORMATTER interpreter routine in the system. The task of this is to convert the ASCII coded data arriving from the interface into binary code, or the binary data to ASCII code.

The data arriving from the interface can gain entry into the computer's memory in one of two ways. One method is the programmed I/O, the other is the autonomous I/O. The basic difference between the two methods is that in the case of programmed I/O the data transfer is executed byte-by-byte, while using autonomous transfer, data parcels are transferred.

The transfer time between the computer and the devices has the following constituents:

$$t = t_E + t_T n + t_K$$

where t is the transfer time of the whole message, t_E is preparation time, t_T is the transfer time of one byte, n is the number of bytes in the message and t_K is the code-conversion time.

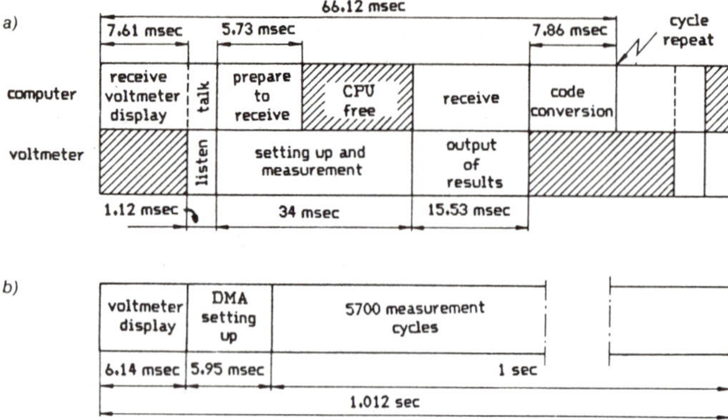

Figure 3.22 Timing of a minicomputer-controlled measuring system. (a) Interrupt I/O, (b) autonomous I/O

Depending on the connected devices and the measuring task, the values of these constituents can vary considerably; however, their numeric values can be determined by empirical measurements.

Figure 3.22(a) shows the timing relationships of the data exchange between an HP 1000-type computer and an HP 3455A digital voltmeter as a typical example. The Type 3455A voltmeter executes 24 measuring cycles per second in the DC operational mode and gives out 14 bytes in every cycle. As is apparent from the Figure, the execution of a complete measuring cycle takes 66.12 mseconds. Out of this time 34 mseconds is allocated to the computer's central unit's use; in the remaining time it can deal with other tasks. The CPU load is thus about 57%.

Figure 3.22(b) shows the timing relationships of an HP 1000 type computer and an HP 3437A high-speed voltmeter connected to it. This instrument, specially developed for automatic measuring systems, can execute 5700 measurement cycles per second, hence the use of the autonomous DMA channel is advisable. The complete measuring cycle shown in the Figure takes 1.012 seconds; the CPU load, including the entire preparation and transfer time, is about 2.2%.

The above examples illustrate that the CPU load is less if DMA transfer is used. In addition, the DMA method is faster if the transferred data parcel consists of more bytes. Although the preparation time is longer using DMA transfer, the total time of the data transfer is shorter (Figure 3.23). For this reason the computer manufacturers always give two values when they give the fastest data transfer speed; one for the autonomous operated DMA channel and the other for programmed I/O.

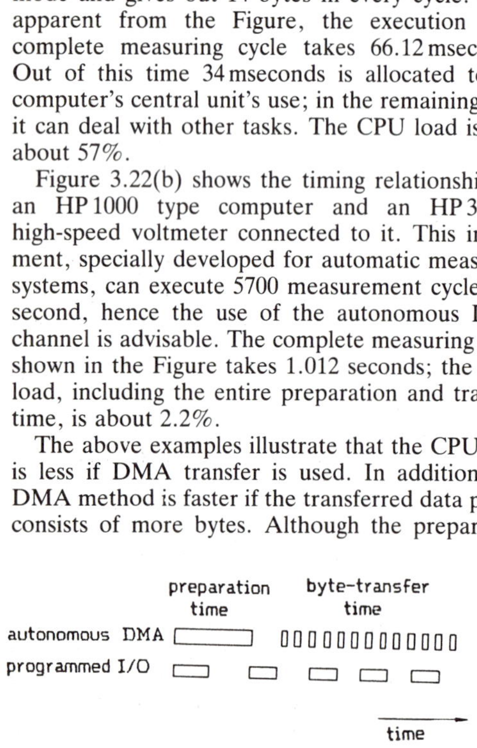

preparation time
byte-transfer time
autonomous DMA
programmed I/O
time

Figure 3.23 Loading of the central unit using interrupt and autonomous I/O

3.3 Microcomputers

Microcomputer manufacturing is the fastest developing area of the semiconductor industry. The number and choice of products available continues to grow markedly. Considering the present development trend of computing technology it seems certain that in a few years microcomputers will play a fundamental and indispensable part in almost every aspect of instrument and measuring technology and automation. In many respects the latest microcomputers already approach, sometimes even surpass, the capacity of minicomputers. Apart from this, they are more advantageous regarding size and cost.

Their use as automatic measuring system controllers is not the only application area for microcomputers. Many more instruments are using built-in microcomputers for the execution of internal control tasks and simple arithmetic operations. These so-called 'intelligent' instruments increase the capacity of the system and simplify the task of automatic system designers.

From the user's point of view the main difference between mini- and microcomputers lies in their relative states of completeness. The minicomputers used in measuring automation are purchased in most cases with completely developed software and hardware (turnkey systems). In contrast, the final development of microcomputers is generally the user's task. Accordingly, the microcomputer user must be more self-reliant and more familiar with the hardware and software structure than a minicomputer user. For this reason the structural elements, as well as their most important characteristics, will be discussed in the following section. In the discussion individual characteristics will not be given, as these can be found in the appropriate instruction manuals. Our purpose is to summarize the most important features systematically.

3.3.1 Classification of microcomputers

Microcomputers can be classified in various ways. The basis of classification can be the word length of the central unit, i.e. the number of bits. This is a very important characteristic and it determines the possible application areas.

4-bit microcomputers are used mainly in calculators and cash registers, although the programmed controller application – previously the application of traditional hard-wired digital circuits – is also an important application area.

8-bit general purpose microcomputers are extensively used to control intelligent peripherals and instruments.

16-bit microcomputers are used in areas requiring high accuracy and fast operation. The resolution, determined by the word length of the computer, is an important characteristic in measurement technology applications. If the sign of the data to be processed is marked by 1 bit, then a 4-bit word length provides 13% resolution; whilst the resolution is 1% using an 8-bit word length and is 0.003% using a 16-bit word length.

32-bit microcomputers are suitable for complex systems containing multiple central processors, DMA transfers and high-speed graphics. An important feature of their development is the integration of the latest improvements in hardware with the new methodology of software engineering.

Alternatively, microcomputers can be classified according to the structure of the central unit. From this aspect they can be divided into four groups:

(a) *Standard structure* (Figure 3.24(a)). Here a microprocessing unit (MPU) plays the role of the traditional central unit. The memories and the input/output unit (I/O) are contained in another chip and are connected with the microprocessor via the data bus and the address bus. The bus driver, the clock signal generator and its controlling crystal are generally external units.
(b) *Single-chip microcomputer* (μC) (Figure 3.24 (b)). In this case all units of the microcomputer are contained in a single semiconductor chip. An important characteristic of this structure is that it has at least three I/O channels and the storage capacity is very limited.
(c) *Multi-chip computer*. The structure shown in Figure 3.24(c) is the most generally used of the many variants in practice. One of the integrated circuit chips contains – apart from the microprocessor elements – the clock-signal generator, the bus driver and part of the I/O unit. The other chips contain the memories and the I/O units.
(d) *Bit-sliced structure* (Figure 3.24(d)). In this modular-type structure the central unit and the corresponding microcomputer characteristics are developed by the user according to his needs.

3.3.2 The operation of the central unit

The structure of various minicomputer central units is very different. However, some basic functional units can be found in all the versions.

The task of the *arithmetic and logic unit* (ALU) is the execution of basic operations on the data. These operations are:

- binary addition,
- Boolean algebraic operation,
- complement generation,
- contents increment/decrement, and
- stepping.

All complex tasks to be solved by computers are based essentially on these operations.

In the arithmetic unit there are some bi-stable storage units as well. The task of these is the storage of important characteristics of the operations executed in the arithmetic unit.

The *zero flag* signifies that the result of the operation is zero.

The *sign flag* indicates the sign of the result.

The *carry flag* signifies that a carry-over is to take place between the bytes of a multiple-byte data string.

The overflow flag indicates overflow during the executed operation.

Other important activites of the central unit are the control and timing of the arithmetic unit and the operation of loading into memory.

Apart from these two important sub-elements, the central unit also contains various temporary storages, the so-called 'registers'. These are the following:

(1) *Accumulator* (ACC): During arithmetic operations the operand is stored in this register and the result will be stored here after the operation has been completed. Apart from this, it is connected to the data bus and serves as the data source during the process of writing into memory or during data output. All data go to

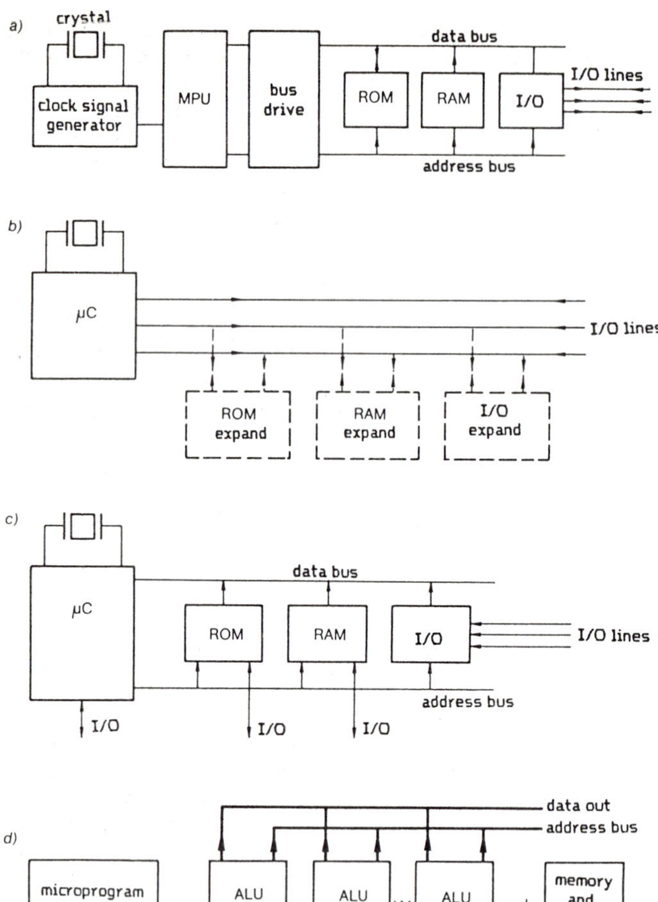

Figure 3.24 Structure of microcomputers. (a) Standard structure, (b) single-chip, (c) multiple-chip, (d) bit-sliced

the accumulator during the process of reading from memory or during data input. The bit length of the accumulator is identical to the number of bits of the central unit's word length. As most microcomputers have 8-bit word length, 8-bit accumulators will be assumed when the operation of the central unit is discussed.

(2) *Program counter* (PC): This register stores the storage address that is to be used for reading the next instruction. The contents of the program counter is incremented by 1 after the execution of each instruction and can be modified using certain commands such as the various conditional and unconditional jumps, overwriting of the program counter contents, etc. These are used in program branching and in subroutine handling.

(3) *Address register* (AR): In the case of storage referred instructions the address register contains the memory address of the operand (data). The word lengths of both the program counter and the address register depend on the size of the memory to be addressed directly. In practice both registers are of 16 bits, allowing the direct addressing of a 65536 (64K) word memory.

(4) *Instruction register* (IR): Its task is the storage of the coded instruction arriving from the addressed memory compartment.

(5) *Free registers:* The instruction system of every microcomputer contains commands that allow

the direct execution of operations using the contents of the microcomputer's internal registers. This greatly reduces the operational time, because the operations of reading from and writing into memory are omitted. The free registers are therefore used mainly for storing constants and intermediate results.

(6) *Stack memory* is a register group whose members are interconnected in such a way that the reading of their contents is possible only in reversed order to the writing. Only the first register is available for reading or writing. Prior to writing, the contents of all registers are transferred to the next one, thus allowing writing to the first one. When the contents of the first register are read, the contents of the other registers drop by 1 to refill the first one again. The stack memory has great importance in handling operations with subroutines, among other activities. If a program to be continued later is temporarily abandoned, then the contents of the program counter and some registers must be saved. The stack memory can be used for this purpose, as its operation ensures that return to the program that was interrupted occurs in every case.

The stack memory is not always located in the central unit, but is often to be found in the data memory. In such cases a *stack pointer* (SP) is put into the central unit. The contents of the stack pointer are incremented by 1 when writing into the stack register and decremented by 1 when reading from it occurs.

Figure 3.25 shows the structure of a microcomputer's central unit using the above-mentioned elements. The Figure is simplifed on purpose, omitting the bus lines leading to the external units. The sequence of all operations is directed by the control unit (CU). The operation of the central unit is, however, controlled by the contents of the instruction register. The structure of the individual microcomputer central units is much more complex than this and is characterized by great variety, in order to suit different applications.

3.3.2.1 Basic instruction cycle

During the operation of the computer the data transfer between the individual units takes place on the so-called 'buses'. The bus is a group of the signal lines that allow information transfer in parallel binary form (in bytes). More than one unit can send signals to any one bus, but on any one bus only one byte can be transferred at any one time. For instance, it is not possible that a data and program memory connected to a common bus should send data simultaneously to the central unit.

The data transfer operations and the cyclic operation of the computer's central unit is timed by a crystal-controlled clock signal generator which may have several phases. The frequency of the clock signal is typically between 1 and 10 MHz.

The execution of every instruction occurs in two stages in the computer:

(a) During *instruction fetch* the central unit switches the contents of the program counter to the address bus. This and the corresponding control signals allocate the next instruction to be executed by the central unit. The allocated instruction is transferred to the central unit's instruction register via the data bus and the internal control of the central unit simultaneously increments the contents of the program counter by 1. Thus it will store the address of the next instruction ready for the next stage.

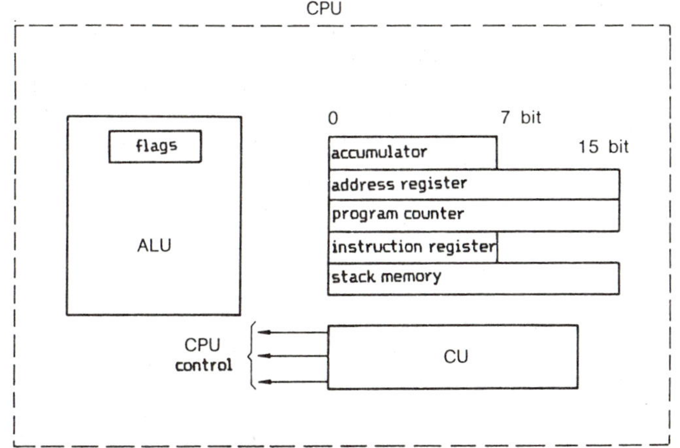

Figure 3.25 Structure of the central unit of a microcomputer

(b) The second stage, the *instruction execute,* begins when the instruction is in the instruction register. This will start the central unit's control circuits, where the execution of the instruction will take place (Figure 3.26.).

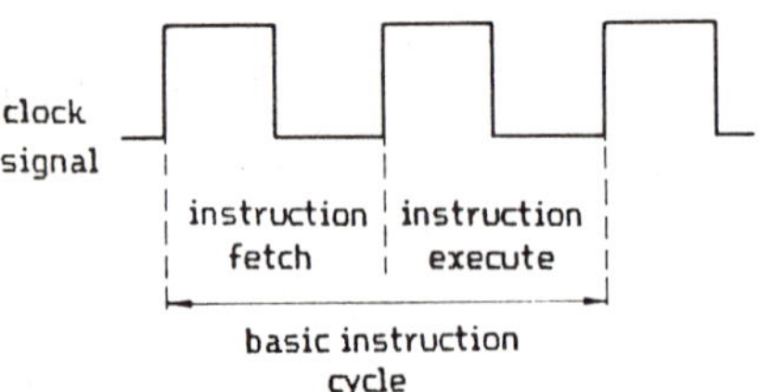

Figure 3.26 Basic instruction cycle of microcomputers

The above process, an entire cycle of the information exchange between the computer's central unit and the memory, is called the 'basic instruction cycle'. The time of the basic instruction cycle is twice that of the clock-signal period.

Let us examine the *timing relationships* of the processes taking place during the basic instruction cycle in more detail. To do this, the hardware structure of the central unit must be known. The central units of modern microcomputers, the microprocessors are marketed generally in 40-pin DIP (dual in-line package) integrated circuits, using LSI (large-scale integrated) technology. Figure 3.27 shows typical IC pin allocations of a microprocessor. This theoretical model is very suitable to demonstrate the operation, owing to its simplicity and clarity.

The first four pins of the microprocessor provide the power supply, earth and clock-signal connections. The next 16 pins (A0...A15) are used for addressing the memory. The two-way data traffic between the central unit and memories takes place on the eight data pins (D0...D7). The signals appearing on the read (R) and write (W) pins have a particularly important role in the computer's operation. During the execution of the program instructions the central unit carries out two basic operations: read from memory and write into memory. Apart from some additional control signals, these basic operations are controlled by the read and write signals.

Let us examine this in more detail, considering the timings. Figure 3.28 shows the timing relationships of the simplest instruction, the 'read memory'. As in every instruction cycle, the start of the operation is the instruction fetch. At the end of the fetch cycle the coded format message is transmitted from the memory to the data bus. The message is decoded and interpreted by the central unit, following which the execution occurs – in our example, the fetching of data from memory. As is apparent from the Figure, the output control signals of the control signals are the same for both data and instruction fetch.

However, there are two significant differences between the instruction fetch and the execution cycle:

(1) In the instruction fetch cycle the memory address given to the address bus originates from the program counter's register, while in the execution cycle the contents of the address register are transferred to the address bus.

(2) In the instruction fetch cycle the contents of the data bus are transferred to the instruction register, while in the execution cycle the data are sent to the accumulator.

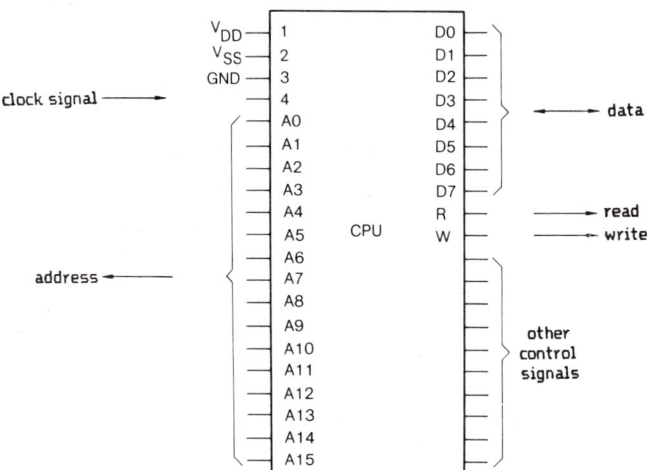

Figure 3.27 Pin allocation of microprocessors (theoretical model)

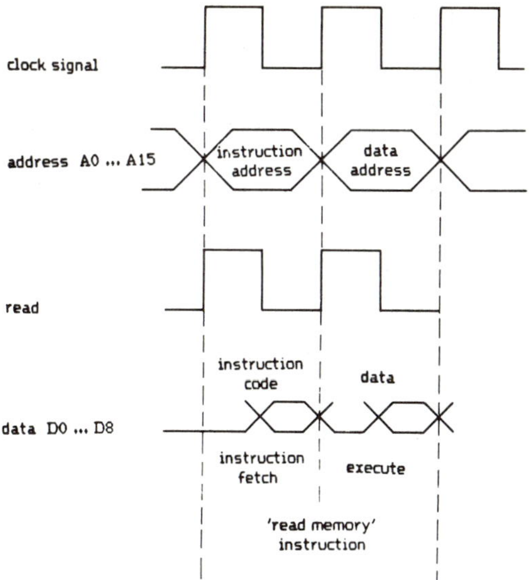

Figure 3.28 Timing of the 'read memory' command

The execution of the operation described in our example is a comparatively simple task. If both the read signal and the clock signal are H level, then the memory circuits must decode the information arriving on the A0...A15 address bus. Following this, the selected memory word – produced by the L level of the clock signal – must be transferred to the central unit.

Using the previous example, let us study the 'addition' instruction. The first stage of the instruction cycle – the fetching of the allocated memory word – is identical to that which would take place with a 'read memory' instruction. The difference is that with 'addition' the fetched data are added to the contents of the accumulator, while with 'read memory' they are transferred to the accumulator unchanged.

The situation is somewhat different with the 'write memory'. In this case the task to be executed is that the data contained in the accumulator are to be written to an allocated address of the memory. This address is stored by the address register. The execution of the 'write memory' instruction can be seen on the timing diagram shown in Figure 3.29. Compared to the two former instructions, the basic difference is that the operation of the memory is controlled by the 'write' signal of the central unit in this case.

3.3.2.2 *Multi-cycle instructions*

The capacity of computers is greatly influenced by their instruction sets. In the above example an 8-bit data bus and a 16-bit address bus were assumed. With this generally used arrangement only the simplest instructions can be executed in a single cycle. The transfers between registers, stepping, incrementing and decrementing are such instructions.

The majority of instructions consist of several bytes, called 'multiple-byte instructions'. This means that the complete instruction code is a multiple of the byte's bit number, in our example consisting of 16, 24, 36, etc, bits. The individual bytes of the complete instruction code are located in consecutive memory addresses.

In handling multiple-byte instructions it is necessary that the central unit should begin the execution after the last byte has been fetched. The signal required for this is contained in the first byte of the multiple-byte instruction. To distinguish this code, the name *operation code* (opcode) is used.

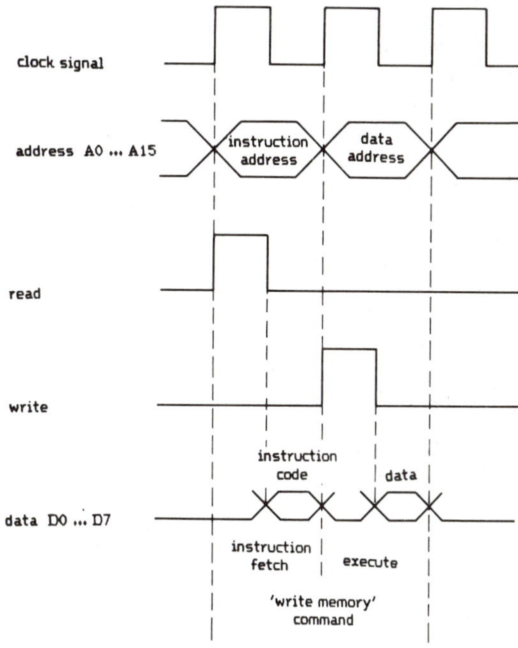

Figure 3.29 Timing of the 'write memory' command

Let us examine the execution of a multiple-cycle instruction using an example involving the Intel Type MCS-85 microcomputer. Our chosen example is the *storage of accumulator contents* (STA). During the execution of this instruction the contents of the accumulator must be written into the given memory address. The selected address is contained in the second and third byte of the instruction:

- 00110010 operation code (STA);
- 10101101 memory address, first byte;
- 10111111 memory address, second byte.

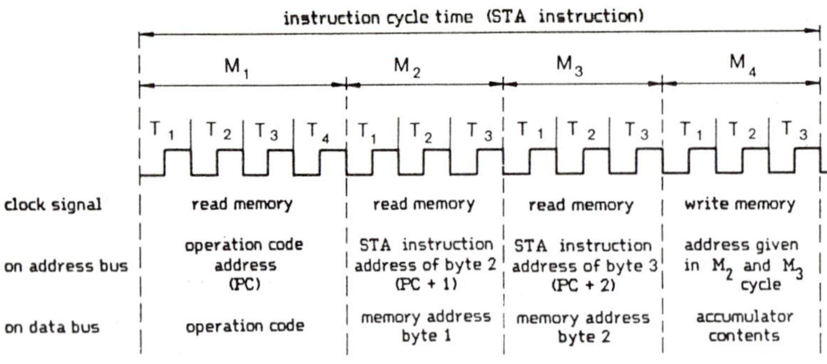

Figure 3.30 Execution of a multiple-byte instruction

Figure 3.30 shows the timing relationships of the STA instruction. During the first machine cycle (M_1) the central unit transmits the contents of the program counter to the address bus and the operation code fetch is executed by the 'read memory' cycle. During the last period of the M_1 cycle's clock signal the central unit decodes and interprets the operation code. After the recognition of the STA instruction the central unit's control circuits are already 'aware' that another three machine cycles will be necessary to execute the instruction. Such multiple-byte instructions are stored in consecutive addresses of the program memory, thus their addressing is ensured by the simple stepping of the program counter. This is indeed the situation in our example also; on the PC + 1 address the central unit receives in the 'read memory' cycle the first byte of the memory address that is to be used for writing the accumulator contents into. The second byte of the memory address is transferred to the central unit in a similar way. Thus the central unit receives all three bytes of the STA instruction, following which the execution can commence. In the fourth cycle (M_4) a 'write memory' cycle transfers the accumulator contents to the allocated memory address. Then the execution of the program continues, the central unit beginning the next instruction fetch.

According to the above, it is apparent that the time necessary for the execution of the instructions, the so-called 'instruction cycle time' is dependent upon two factors. One of these is the magnitude of the clock-signal-dependent basic cycle time of the central unit, whilst the other is the number of basic instruction cycles necessary for the execution of the instruction.

3.3.2.3 *Instruction set*

One of the most important characteristics of a microcomputer is the instruction set available for interpretation by its central unit. The number of instructions available, although important, is not the primary factor in determining the usefulness of the instruction set. The efficiency of the instructions is a more important characteristic, as it determines the size of the memory required for the execution of a given program. This is highly relevant in that the cost of memory forms a significant part of the total cost of the computer.

More importantly, the instruction set greatly influences the operational speed of the computer and this is a highly important aspect in measuring technology applications.

The instruction set of microcomputers is determined by the structure of the central unit and by the number of registers and their mode of addressing. As the structure of the central unit of different microcomputers varies greatly, the instruction sets also are varied in their contents. However, there are four instruction groups that are to be found in all types of microcomputer and which can be separated in some cases. These are as follows:

(1) *Arithmetic and logic instructions:* The various operations executed by the arithmetic logic unit (ALU) belong to this group. Operations requiring two operands are executed using data stored usually in the accumulator and in one of the registers, maybe at a given memory address. The various flags play an important role with these instructions. The different types of flag were listed when the structure of the central unit was discussed.

Most microcomputers use binary arithmetic and two's complement coded number representation and it is worth noting that most types are suitable for handling BCD coded data.

(2) *Data transfer instructions:* The instructions connected with the data stored in the program or data memory are in this group. The STA instruction mentioned in the previous chapter is one of these, for example. The data transfer instructions often contain arithmetic or logic tasks too.

(3) *Control instructions:* The control instructions have a major role in computer programming. The various jump and branching instructions, the skip instructions, subroutine call and return make it possible that the central unit can interrupt the stepping sequence and freely access between certain program segments during the execution of the program.

The control instructions can be either unconditional, i.e. always to be executed; or conditional, when the jump or the branching depends on a condition. Usually the condition is a defined state of a given flag.

(4) *Input/output instructions:* The computer's central unit and the memory are connected to the external peripheral units va the input/output unit. The I/O instructions handle the data traffic concerning these peripherals. Hence, these instructions will determine the way the user can obtain data from the computer and the data input method.

In most cases the above four instruction groups cannot be divided in a clear and unambiguous manner, as several instructions can also be classified in two or three groups.

There are some important special instructions that do not belong to any of these groups. The HALT instruction available in every microcomputer is an example of such instructions.

3.3.3 The memory

The memory of a digital computer is a series of binary storage elements organized in a defined system. According to the principle of the storage, there are two different types of memory:

- magnetic memory, and
- semiconductor memory.

Magnetic memory is provided by ferrite-core magnetic-storage elements and these have been used extensively in mainframe computer systems. They are now often replaced by semiconductor memory and will therefore not be considered further.

Semiconductor memory has been used for storing programs and data in microcomputers from the very outset. The user can readily assemble storage capacity appropriate to his requirements. Additionally, semiconductor memory is of course directly compatible with microprocessors manufactured using the same technology.

3.3.3.1 Random-access memory (RAM)

The family tree of the semiconductor memories is shown in Figure 3.31. One large group is the direct access read/write memory (random-access memory, RAM). The contents of RAM can be altered using external control. RAM are manufactured using bipolar or MOS technology.

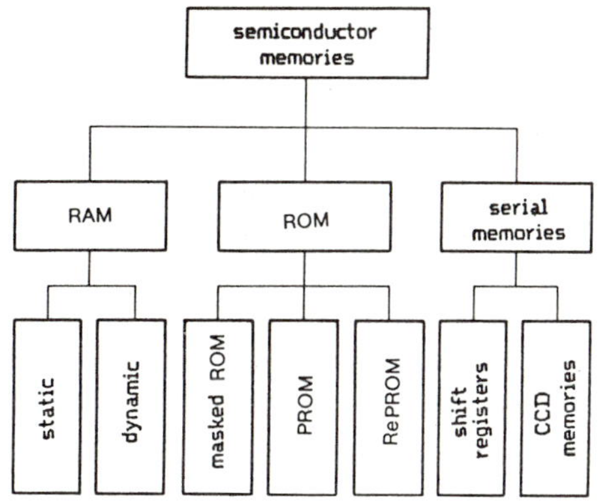

Figure 3.31 Classification of semiconductor memories

The storage cells of a bipolar RAM are of static operation, i.e. they keep the information only while they have a power supply.

The MOS memory cells are of either static or dynamic operation. The dynamic storage cells store information for only a short time, even when the power supply is maintained, and updating or refreshing is necessary every few milliseconds for continuous storage. Dynamic RAM memories have two significant advantages over static memories:

(1) They require considerably reduced power input in their rest state.
(2) The dynamic memory cell contains fewer transistors and therefore a much larger capacity memory can be built up on a given size semiconductor wafer.

At present MOS technology dominates semiconductor RAM manufacture, and dynamic memories are cheaper than static ones. This is counterbalanced by the additional cost and corresponding design problems of the update/refresh circuits. Overall, the use of static RAM is more suitable in small-capacity memories, while it is more economic to use dynamic MOS memory cells for medium- to large-capacity memories.

The bipolar memories (LS, TTL, ECL) are much faster in operation, but the storage capacity contained in a given size wafer is much less. Due to the considerably bigger power consumption and higher costs they are used predominantly in high-speed (e.g. bit-sliced) microcomputers.

Semiconductor RAM are LSI circuits, containing the storage cell block (memory matrix), address decoders, read ampliers and the control logic on a single chip. The addressing mode of the memory is determined by the internal structure.

Figure 3.32 shows the internal structure of a 4096-bit capacity dynamic RAM (Intel, Type 2107B). The memory matrix consists of 64 rows and 64 columns. The selection of the required cell is

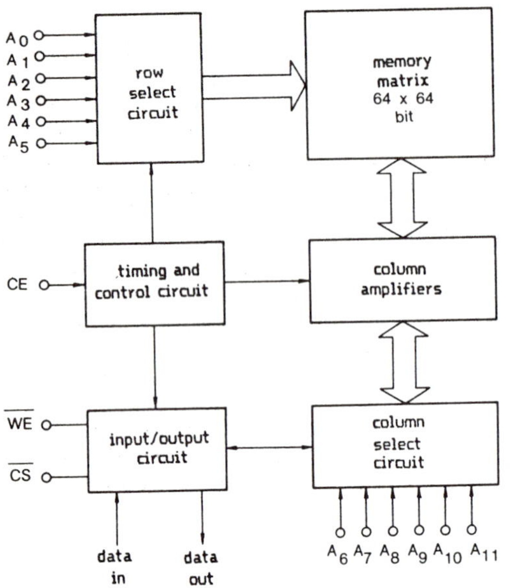

Figure 3.32 Internal structure of the Type 2107B RAM made by Intel

executed by assigning the $A_0 \ldots A_5$ row addresses and the $A_6 \ldots A_{11}$ column addresses. During the operation of the memory all operations are started by the 'chip enable' (CE) signal and timed by the timing and control circuit. During the write operation the input will be connected to the required cell as a result of the 'write enable' (WE) signal. The read operation consists of two cycles. In the first cycle the contents of the addressed storage cell appears on the data output, while in the second cycle the information lost during reading is rewritten in the cell. The so-called 'chip select signal' (CS) – that controls the three-state data amplifiers – has an important role in the operation of the memory, making it possible for the user to build up a memory best suited to his requirements from several RAM chips. In our example the chip select signal originates from an external decoder, although RAM units containing the CS producing logic are also available.

3.3.3.2 Read-only memory (ROM)

The read-only memory is a storage unit with pre-determined contents which can be filled (i.e. programmed) either during the manufacture of the memory (mask-programmed ROM) or by the user (programmable ROM, PROM). Certain types of read-only memory can be cleared, then reprogrammed (RePROM or EPROM).

The use of mask-programmed ROM presents some limitations. The production of the individual masks leads to additional preparation tasks – and hence costs – for the manufacturer. Accordingly, they undertake the production of masked ROM only above a certain quantity (typically 1000). The production of the individual mask is a comparatively long process of some weeks' duration. If the designer makes an error in the original specification of the mask pattern, then the entire production of ROM manufactured becomes useless, as there is no possibility of reprogramming the masked ROM. Therefore, masked-ROM units are generally used only in tested products to be manufactured on a large scale.

In microcomputer development and in products manufactured on a small scale the user-programmable PROM or EPROM memory units are used. The programming of these is executed on the chip.

The programming of PROM is done either by blowing the fuses belonging to the individual memory cells or by the breakdown of reversed biased storage diodes, according to the type of PROM. This process is irreversible, i.e. the reprogramming of the memory is impossible.

As opposed to this, the memory units manufactured using the MOS technology can be reprogrammed by the user, should the need arise. One manufacturing method of the erasable and reprogrammable memory units is the FAMOS process, patented by Intel. The erasure of the FAMOS cell (EPROM) is carried out using high-intensity ultraviolet (UV) light, a transparent quartz cover placed above the wafer ensuring the UV light transmission. However, the entire memory life is erased with this technique.

Another manufacturing process uses metal nitride oxide semiconductors (MNOS) for the storage elements of the basic cells. This method permits selective cell erasure, (i.e. only the memory section that is to be modified need to be cleared), by applying a high-voltage pulse to the appropriate programming pins. The device is known as the 'Electrically-Alterable ROM' (EAROM).

3.3.3.3 Microcomputer memory units

In Figure 3.33 the microcomputer's central unit and the memory units connected to it are shown, with the data and address buses providing the connection. The central unit is controlled by the instruc-

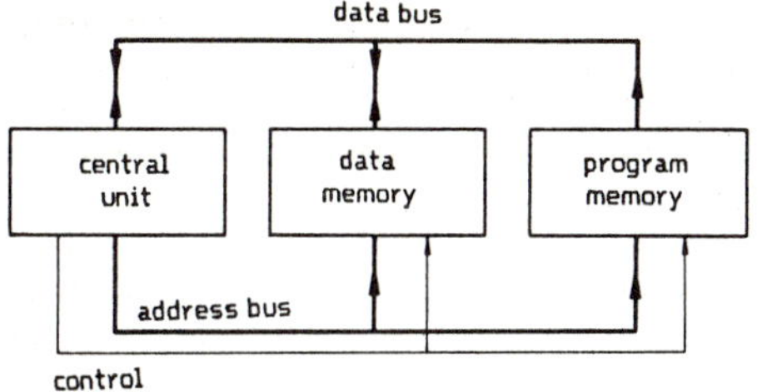

Figure 3.33 Connection of the microcomputer's memory and the central unit

tions stored in the program memory. The central unit executes the operations prescribed in the program using the data stored in the data memory.

The capacity of the program memory is usually between 512 bytes and 32 kbytes and its access time has to suit the operational speed of the central unit. The use of ROM, PROM or EPROM units as program memory depends on certain other aspects such as the quantity of systems built, etc.

The requirements concerning the data memory (e.g. storage capacity, operational speed) are determined by the nature of the application. If fast operation is of primary importance, then bipolar or static MOS-RAM units should be used, while dynamic MOS-RAM units are best when large storage capacity or low power consumption are more important factors.

If large storage capacity (16–64 kbytes) is needed, the data memory usually contains several RAM-integrated circuits. These operate in parallel mode, every chip storing one or more bits of the complete memory word. In such cases all RAM chips receive identical address signals, but their outputs are connected to separate data lines.

Apart from the data and program memory, the microcomputer contains other memory units as well. The register block closely co-operating with the arithmetic unit was described in detail when the central unit's operation was discussed. These registers are basically fast-operating memory units whose task is the temporary storage of the data and addresses being processed.

3.3.4 I/O units

Manufacturers build up families of peripheral devices around their microprocessor units. A family usually includes RAM, ROM and PROM units of various capacities and a number of I/O units, as well as the central unit. These families can be extended in due course, as the manufacturer develops new units, keeping up with technological advancements. It is desirable and often the case, that the elements of a new microcomputer family are compatible with an earlier central unit of the manufacturer, or conversely, the former auxiliary units can be interfaced to the new central unit.

There are units that can be found in almost every microcomputer manufacturer's program, a typical example being the Universal Asynchronous Receiver Transmitter (UART), which is an I/O interface executing the parallel/serial conversion of data.

3.3.4.1 Serial I/O interfacing

The UART units, manufactured by several companies were the first standard LSI circuits in the industry. Although more advanced UART units have been marketed recently by semiconductor manufacturers, the principle remains basically unchanged.

The UART unit consists of three parts: the receiver, the transmitter, and the controller (Figure 3.34). The serial data to be converted and the clock

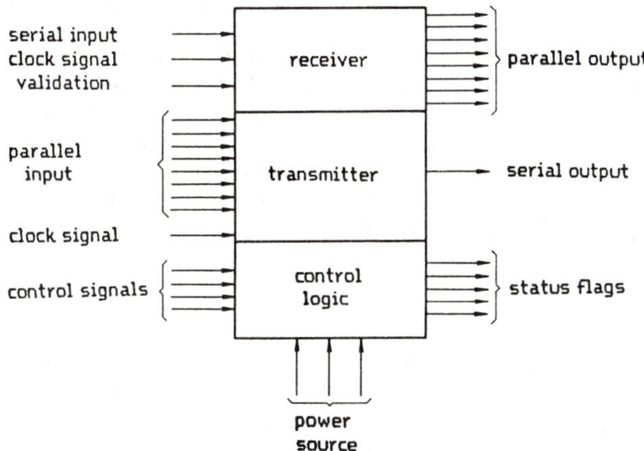

Figure 3.34 Structure of the UART units

signal are transferred to the receiver, producing 8-bit parallel data on the output. The transmitter executes the reverse operation simultaneously, by converting 8-bit parallel data into serial data. The controller receives control signals from the microprocessor and transmits state signals to it. Apart from the serial/parallel conversion the UART unit handles some other auxiliary functions as well (e.g. start and stop bit generation, parity check, etc).

UART units are used for data transfer executed by slow serial peripherals permitting a maximum speed of 9.6 kbaud.

Another, considerably faster, serial interface unit is the USRT (Universal Synchronous Receiver Transmitter). It differs from the UART in that it uses a synchronizing input signal, as well as the serial input and clock signals. The USRT unit is used for fast transfer, for example in the traffic between the microcomputer and the data-transfer modems. The operational speed of these units can reach 500 kbaud.

Single chip UART/USRT units are also manufactured, such as the Intel Type 8251.

3.3.4.2 Parallel I/O interfacing

The fast peripherals and the various display units are interfaced to the microcomputer using parallel I/O connection. Parallel data transfer is used generally for interfacing instruments as well as A/D and D/A converters.

The functions of the individual signals are user defined by programming the control logic. The programmability of the data lines refers to their direction. According to the programming, the same port can be either an input or an output port.

The data on the data bus of the microprocessor are valid for the duration of only a few clock signals (e.g. 500 nseconds for the Intel 8080 microprocessor). The I/O device, on the other hand operates asynchronously (like the clock signal) and is capable of receiving the data only after a few mseconds. Therefore the interface unit, on receiving the write (WR) control signal of the central unit's output command, takes a sample from the databit and stores and transmits it to the I/O device.

The I/O device signals the parallel interface, usually with a handshake control, that it has data on its output. This is signalled by the interface to the central unit using an interrupt; the central unit then jumps to the read I/O program. When the read (RD) control signal of the program's output command is received, the parallel interface gates the data onto the data bus. Following this, using a signal transmitted with the output command, it signals the I/O device that it had received the data and is now ready to accept new data.

3.3.4.3 Analogue I/O units

In the introduction it was mentioned that the majority of physical quantities are analogue, i.e. continuously-changing in time. Hence, certain I/O units of microcomputers used in measuring technology, the analogue/digital (A/D) and the digital/analogue (D/A) converters, are very important. These units convert the analogue signal into a digital form acceptable to the microcomputer; or conversely, the digital output signal of the microcomputer into an analogue quantity.

Some special aspects have to be considered in designing these units:

(1) The microcomputer operates at a speed determined by its internal clock signal which is entirely independent of the speed of the analogue signal.
(2) The majority of microcomputers provide 8-bit resolution. This corresponds to approximately 0.8% resolution of the analogue signal, which is not adequate in every case.
(3) The operations executed by software are comparatively slow in microcomputers, therefore the majority of these should be executed by the hardware in the I/O units.

A general aim is to reduce the number of integrated circuit chips constituting the system. This increases the reliability and decreases costs.

The converters designed satisfying the above aspects contain memory, gate and analogue switching circuits in addition to the traditional A/D and D/A circuits (Figure 3.35).

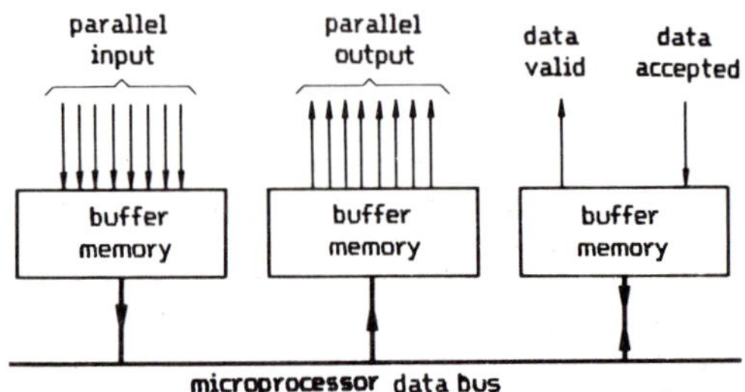

Figure 3.35 Parallel I/O connection via gated buffer memories

In the D/A converters the information appearing on the data bus of the microprocessor is transmitted to the buffer which will save it until the next 'update'. This ensures the continuity of the analogue output signal of the converter. In the A/D converters the role of the buffer is the saving of the address signals that control the operation of the input multiplexer.

If the converter has 10, 12 or 16 bits – to achieve better resolution – an additional problem arises in

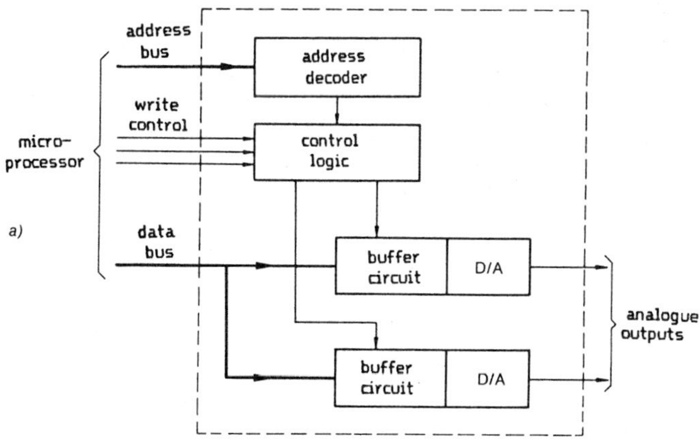

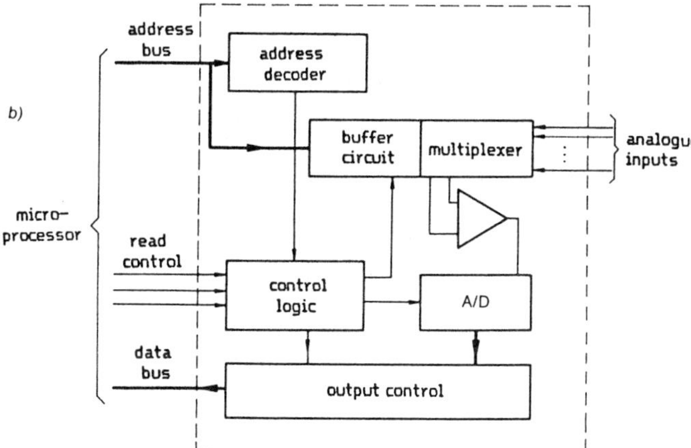

Figure 3.36 Structure of the Burr-Brown analogue I/O units. (a) Type MP 10/11, (b) Type MP 20/21

that the converter cannot be interfaced directly to the 8-bit data bus of the microcomputer. The data must be split for transfer between the two systems and therefore the transfer usually takes place in the form of two consecutive bytes. This complicates the internal control of the converter and slows down its operation.

Figure 3.36 shows the block diagrams of the analogue I/O units made by Burr-Brown. In the Type MP10/11 analogue output unit shown in Figure 3.36(a) two 8-bit D/A converters and all auxiliary logic units necessary for interfacing the micro-computer can be found. The Type MP20/21 analogue input unit shown in Figure 3.36(b) is a complete data collecting system suitable for digitizing 16 analogue input signals.

These 8-bit I/O units can be controlled very simply. From the microprocessor's point of view the analogue output unit takes up the role of an 8-bit register and new data can thus be given to the output using only one instruction. The analogue input can be controlled by the program similarly; for example, in case of the Intel 8080A a single LDA instruction is sufficient for the input of an analogue channel. If necessary, the latter unit can be interfaced to the microcomputer using the interrupt principle.

Analogue I/O units were first manufactured using hybrid technology and in consequence were quite expensive. Using the combination of the I^2L (Integrated Injection Logic) and CMOS technologies the manufacture of cheap monolithic converters has become possible.

3.3.4.4 *IEC interface units*

During the development of specially designed IEC systems that use microcomputers as controllers considerable problems are often encountered by the user in developing the software. Hardware design is comparatively simple, as the operation of microcomputers and the IEC system has many common features. This can be illustrated by examining a few important points.

The simplicity of the interfacing is very dependent on the organization of the two systems' data traffic. The IEC system, and the majority of the microcomputers, transfer information on an 8-bit parallel data bus. The arithmetic and logic operations are executed by the microprocessors in 8-bit words and therefore decoding and generation of the 8-bit IEC messages does not present a special problem.

The multiple-byte messages, commonly used in automatic measuring systems, can be generated or decoded quite simply by the microprocessor. The transfer of the data can take place in ASCII code or in simple binary form, since the data-handling ability of the microprocessors makes the rapid conversion between the different codes possible.

The asynchronous operational mode of the IEC bus is a distinct advantage in microcomputer/instrument interfacing, because the microcomputers are generally faster than the other devices comprising the IEC system. Furthermore, in the time interval between individual control cycles the microcomputer can readily deal with other problems, such as calculating or averaging the measurement values. The service request of the IEC system devices can be validated in the form of an interrupt.

The microcomputers can be interfaced to the IEC system using one of several methods. In selecting the method the operational speed and the cost are the two primary aspects requiring consideration.

The required operational speed is determined by the characteristics of the system to be controlled. A data transfer requiring the maximum operational speed of the IEC system (1 Mbyte/second) needs to be considered only if the system contains magnetic disk memory. To achieve the fast operational speed necessary, there is a need for large buffers and for costly circuits (DMA) in the interface. However, the operational speed of the majority of instruments is limited to a few kbytes/second, owing to internal, measurement-dependent, characteristics.

The interface unit also depends on the method used for realizing the interface functions. The state diagram description of the IEC system makes both software and hardware type interfacing feasible.

3.3.4.5 *Software-type interfacing*

For software-type interfacing a universal control unit and a program memory controlling its operation are required. Figure 3.37 shows the block diagram of a software-orientated interface unit. The American ZiaTech Type ZT80 connects the Intel Type

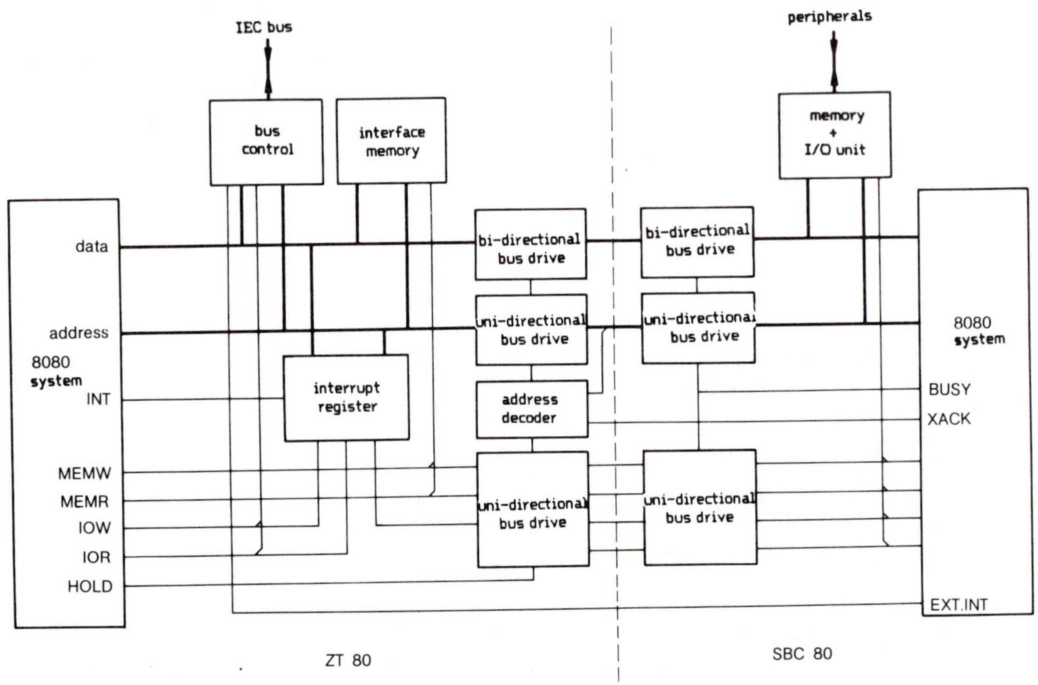

Figure 3.37 Structure of a software-orientated IEC interface (ZiaTech Type ZT80)

SBC80 microcomputer family and the IEC bus. The two Type 8080 microprocessor systems are connected via uni-directional or bi-directional bus driver circuits. The two data buses are connected with a bi-directional, the address bus with a uni-directional bus driven circuit. The messages going to, or coming from, the IEC bus pass through the interface memory which can be accessed from either processor.

In this solution the interface functions are handled independently by the microprocessor contained in the interface unit, and therefore do not add to the load of the microcomputer.

Another advantage of the software-orientated interfacing is that possible changes of the interface requirements can be accommodated by a simple ROM change. The hardware cost is much less than the software cost in this arrangement and this means that this solution can be economical only in large-scale production.

3.3.4.6 Hardware-type interfacing

Hardware-orientated interfacing of a microcomputer and the IEC bus can be achieved in several ways. The interface functions can be realized using SSI and MSI circuits. In such cases a microcomputer/IEC interface can be constructed from 30 to 150 chips, according to the required structure, on a printed circuit card. This solution is very expensive, owing to the design work and to the large number of components.

Hardware-type interfacing can also be achieved by using programmable universal logic modules (Field Programmable Logic Array, FPLA). FPLA circuits are basically LSI units containing AND/OR gates and inverters that can be programmed by the user, similarly to PROM units. The programming is based upon the required logic functions and, in essence, involves the creation of a connection system of the individual logic elements.

The Type 102 universal interface card of the Mikrologic company of Munich, Germany (Figure 3.38) contains an FLPA circuit that realizes one part of the IEC-bus/microcomputer interface, namely the interface functions. The other part of the interface unit – the circuits realizing the device functions – must be built by the user according to the special application and the microcomputer used. Hence, there is free space for an 18-pin DIP integrated circuit on the printed circuit card. The connection of the circuit designed by the user is effected by either wire wrap, or in the case of large quantities the manufacturer will produce printed circuit cards containing the desired connections.

The block diagram of the interface unit is shown in Figure 3.39. The main units are the address comparator, the address selector, the control unit and the character counter.

The address comparator compares the device address set in the address selector with the address arriving from the bus. The control unit is an Intersil Type 5200 FPLA circuit which handles the allocation of the talker and listener state, the data synchronization and the timing of the handshake process. The character-counter facilitates the generation and decoding of the character strings constituting the interface messages.

The timing diagram of the listener addressed interface is shown in Figure 3.40(a). The character counter circuit is stepped by the CL_{IN} clock signal generated in the control unit. The output signal of

Figure 3.38 Type 102 universal interface card made by Mikrologik

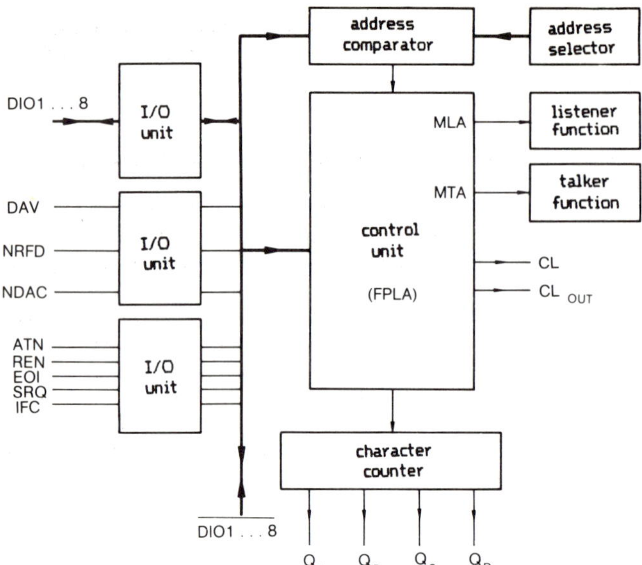

Figure 3.39 Internal structure of the Type 102 Mikrologic universal interface card

the character counter, indicating the location of the characters within the message, and the CL_{IN} clock signal can be used together to control the RAM.

The timing diagram of the talker operative interface unit is shown in Figure 3.40(b). The impulses ensuring the output of the individual characters at equal intervals are generated from the CL_{OUT} clock signal produced in the control unit and from the output signal of the character counter ($T_1 ... T_{n+1}$). The clock signal and the counter allocate the position of the string delimiter (e.g. CR) at the end of the character string.

3.3.4.7 LSI interface circuits

The third, considerably simpler and cheaper method of hardware-type interfacing uses special integrated interface circuits. Two factors relevant in their development were:

- The demand for microcomputer/IEC interfaces reached a level where the semiconductor manufacturers found the development of LSI interface circuits economical.
- Semiconductor technologies (NMOS, CMOS/SOS) developed permitting the manufacture of complex, fast operating, logic circuits whose task is the realization of the IEC control functions.

At present Hewlett-Packard, Intel and other companies produce LSI circuits which provide control functions.

The LSI unit of Hewlett-Packard – manufactured using CMOS/SOS technology – is called 'PHI' (Processor for Hewlett-Packard Interface) and was developed for their MC_2 16-bit word length microprocessor and other Hewlett-Packard instruments. The first combined use of them was in the Type HP 2240A universal measuring and automatic controller.

Intel achieved the entire IEC bus/microcomputer interfacing in two 40-pin integrated DIP chips. The

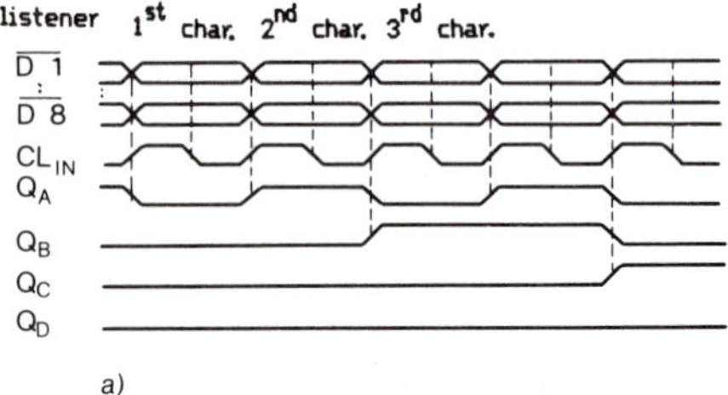

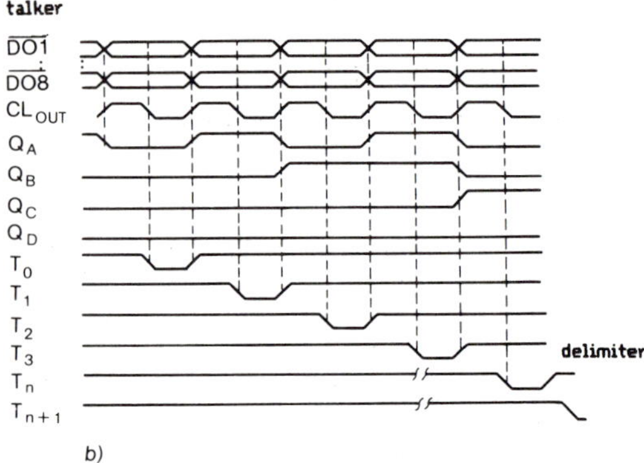

Figure 3.40 Timing of the operations of the Type 102 Mikrologic interface. (a) As listener, (b) as talker

Type 8291 S/790 chip realizes the functions necessary for the operation of the talker and listener devices, whilst the Type 8292 chip contains the control functions. Both circuits can be interfaced directly to every member of the Intel microcomputer family (MCS-80, MCS-85, MCS-48 and MCS-869) and the two circuits are complementary. However, Type 8291 can be used on its own for realizing the handshake functions, service request, local/remote control, parallel poll, device clear and device start, as well as talker and listener functions.

For the use of microcomputers as IEC controllers the combined application of the Type 8291 and Type 8292 interfaces is necessary. The structure of the complete IEC interface is shown in Figure 3.41. The operation of the individual circuits is synchronized by the Type 8292 IEC control circuit. This circuit also generates the ATN, IFC and REN messages, and receives the SRQ message from the bus.

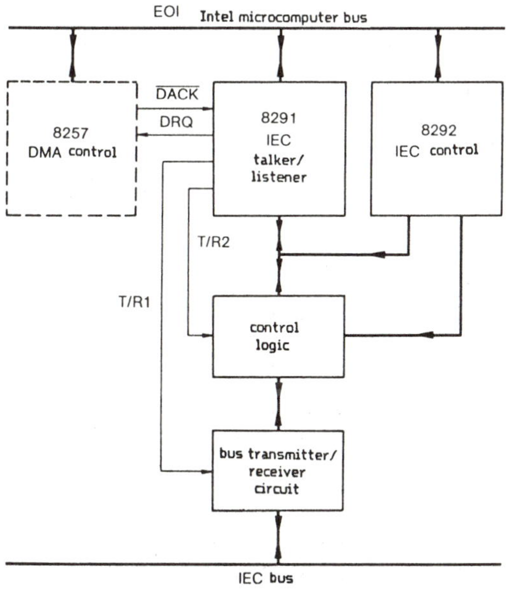

Figure 3.41 Interfacing Intel 8-bit microcomputers to the IEC bus

The Type 8291 unit directly controls the optional DMA unit, the IEC bus side controlling the data traffic control logic and the Type 8293 bus transmitter/receiver units. This unit contains the circuits executing the message coding and address recognition.

The use of the integrated control unit is advantageous economically, as well as providing greater variety and better space utilization. The cost of the Type 8291 and Type 8292 circuits is about one tenth of the cost of traditional hardware or software interfacing.

3.4 Programmable calculators

One of the characteristic features of computing technology is that the difference between calculators and minicomputers is becoming less and less distinct. The modern programmable calculator can be programmed in a high-level program language and its large capacity I/O system permits the interfacing and efficient use of the most varied peripherals. On the other hand, the size of the minicomputer becomes steadily smaller, whilst it contains more and more firmware, and the manufacturers constantly try to build certain simpler peripherals in with the central unit.

The situation is further complicated by the fact that manufacturers do not give a uniform description of their products. It can happen that the same device is called a 'programmable calculator' in one leaflet and a 'desk-top computer' in another. Disregarding these manufacturers' terms, devices that have a strict program language can be programmed in an interactive way, have interface units and are of desk-top format will be dealt with in the following section.

From the user's point of view the most important difference between programmable calculators and computers lies in the method of programming. The programming language of the calculator – determined by the interpreter program stored in the ROM – cannot be changed. The device can be used immediately after it has been switched on and its internal structure is optimized for one programming language only.

Conversely, the computer user has a choice of several programming languages; however, the computer can be used only if the interpreter program corresponding to the chosen programming language has been loaded into the memory.

There are three basic types of calculator programming language. However, the classification cannot be unambiguous in every case, as several devices can be programmed using a hybrid language.

3.4.1 Keystroke language-programmable calculators

The simplest calculator programming language is the keystroke language, which resembles the assembler language of a computer. In this programming language each key corresponds to a given instruction or instruction code. Programming is executed by pressing the required keys in the appropriate sequence and the execution of the program occurs in the same sequence.

The keystroke language-programmable simple calculators do not 'understand' the brackets used in algebraic expressions. For this reason a special programming technique was developed, called 'reverse Polish notation' (RPN), after its inventor, the Polish mathematician Jan Lukasiewicz.

In this system, after keying in the operands, the appropriate operation sign must be keyed in. For example, the expression

$$a + \frac{bc}{d} - e$$

can be entered in the following way:

a b c * d / + e − =

This form is very different from the original expression and the user must become familiar with this unnatural programming technique. An obvious advantage of the system is that the internal operation is very simple, as it executes the operations in the sequence of the keystrokes.

The RPN is not the only keystroke language. The *algebraic-entry system* (AES) and its *hierarchic version* (AESH) are also used.

In the AES system the expression must be rearranged before keying it in can take place. The above example would be executed using the AES system in the following way:

b * c / d + a − e =

In the AESH system it is necessary to observe certain hierarchy in the operational sequence; thus the multiplications and divisions are executed first and there is no need for rearrangement of the expression. The above expression in the AESH system is:

a + b * c / d − e =

Keystroke language-programmable calculators are very useful for solving simple problems, but the creation of long and complex programs is tedious and time-consuming, even for very experienced users.

These calculators can be integrated with certain simple peripherals such as magnetic card readers, printers, etc. A typical device of the keystroke language-programmable calculator family is the Hewlett-Packard Type 97S, shown in Figure 3.42. This device is suitable for processing BCD coded data, using the optional interface located in a separate unit shown next to it.

3.4.2 Algebraic language-programmable calculators

The algebraic (or formula) language-programmable calculators overcome the disadvantages of the keystroke machines, namely the artificial and the difficult programming technique and the poor and limited peripherals. Algebraic languages are a mixture of keystroke languages and high-level languages (FORTRAN, BASIC). As a consequence

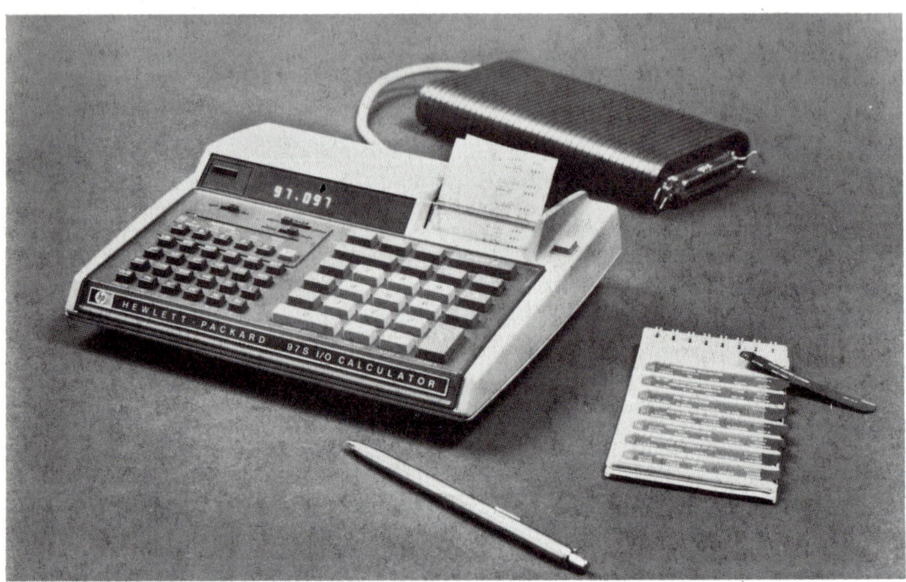

Figure 3.42 Hewlett-Packard Type 97S calculator

of this, the algebraic expressions can be loaded into the machine in their usual form and this makes programming much easier. As an example, the program of the formula for solving a quadratic equation is given in the language of the HP 9820A calculator as follows:

```
 0: ENT "A VALUE",A
 1: PRT "A=",A
 2: ENT "B VALUE",B
 3: PRT "B=",B
 4: ENT "C VALUE",C
 5: PRT "C=",C
 6: IF 4AC>BB;GTO"IMAG"
 7: PRT "REAL ROOTS";SPC 1
 8: PRT (−B+√(BB-4AC))/2A;SPC 1
 9: PRT (−B−√(BB-4AC))/2A;SPC 9;JMP−9
10: "IMAG"
11: PRT "COMPLEX ROOTS";SPC 1
12: PRT "REAL","IMAGINARY";SPC 2
13: PRT−B/2A,√(4AC−BB)/2A;SPC 1
14: PRT−B/2A,−√(4AC−BB)/2A;SPC 9; GTO 0
15: END
```

The first five lines of the program demonstrate the interactive nature of calculator programming. Before the program is run, the calculator asks for the variable values. These variables which are present in the algebraic formula require a data storage field in the machine. It should be noted that in lines 8 and 9 the solution formula is there in its usual form. The calculator interprets the brackets and the hierarchic structure of the formula and executes the operation accordingly.

The algebraic language programmable calculator usually permits program editing in addition to the calculations. Using the editing key the characters present in the individual program lines can be modified and cleared, and new characters can be inserted. The algebraic language calculator recognizes syntax errors of the program and informs the programmer about the nature of the error with an identifying error message.

The programming of calculators is made much easier by user definable keys. The routines corresponding to frequently-used program sections can be stored in the calculator's memory and can be called upon by a single keystroke. Such a routine can be a program line necessary for the automatic calibration of an instrument, or an algorithm calculating a measurement result. The input of the routine is carried out by the programmer, using the keyboard.

The calculator manufacturers market various ROM packed special routine packages solving mathematical or statistical problems, or I/O routines for the control of various peripherals. The use of these routines results in significant saving in memory space requirements.

3.4.3 High-level language-programmable calculators and desk computers

High-level programming languages have two significant advantages. One is that they simplify the programming, transforming the machine–man interaction into a dialogue using alphanumeric messages. The other advantage is that the programmer can use programs in a standard language, written

Figure 3.43 Tektronix Type 4051 graphics calculator

and published by others; furthermore, programs already written and used are not lost when the calculator is changed to another type. The most commonly-used high-level programming language is BASIC.

3.4.3.1 BASIC programming language

The BASIC language in its standard form (ANSI X3.60–1978) contains 64 instructions and owes its popularity to its simplicity, interactive nature and its modular character. The modular character means that the original language, containing only the minimal amount of programming set, can be extended with accessory modules to a programming language suitable for solving more complex problems.

The TEK SPS BASIC, developed by Tektronix for their Type 4051 calculator, is a characteristic example of measuring system orientated, modular BASIC software (Figure 3.43). This device, which is a graphic calculator containing memory display using an Intel 8080A microprocessor, is used by several automatic-measuring-system manufacturers as a control unit. The rapid expansion of the 4051 and other graphic calculators is due to the fact that measurement data displays in graphic form can be evaluated more easily and faster than can number strings or tables containing the same information. The cathode-ray tube display graphic calculators can draw graphs much faster than X-Y plotters and are free from the mechanical problems associated with these devices.

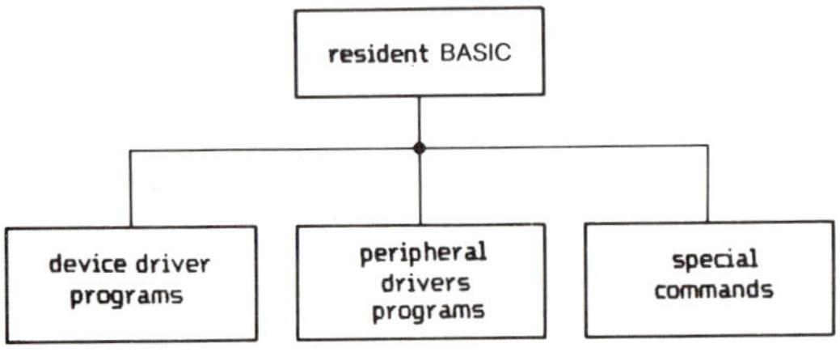

Figure 3.44 Structure of the TEK SPC BASIC software system

The basic version of the Type 4051 contains an 8-kbyte RAM that can be extended to 16 or 32 kbytes, according to the user's need. The programs written in BASIC are translated to the microprocessor's language by an interpreter program stored in a 32-kbyte capacity ROM. On the built-in magnetic tape cassette up to 300 kbyte of data or programs can be stored. The device has a standard IEC interface unit.

The structure of the TEK SPS BASIC modular software system is shown in Figure 3.44. The resident BASIC program, ensuring the operation of the device is kept in the memory permanently. Other parts of the software are transferred into the memory only when they are needed, thus freeing considerable storage capacity for programs and data. The software modules available are the device drivers, the peripheral drivers and the special commands.

The device driver programs are special routines facilitating the control of the IEC interface devices. The peripheral drivers execute the control of the punched-tape readers or line printers, etc, connected to the device via the optional RS-232-C interface. The special commands contain the routines used mainly for graphic data display.

3.4.3.2 HPL programming language

Apart from the standardized and extensively-used BASIC language, there are some other high-level programming languages relating to certain calculator types. One of the best known of these is HPL, the programming language of the Hewlett-Packard Type 9825A calculator.

HPL is a peculiar hybrid of BASIC, FORTRAN, ALGOL and PL/1 languages. The aim in its construction was to minimize the time and memory requirements necessary for programming complex algebraic expressions.

The basic unit of the HPL language is the so-called 'statement'. More than one statement can be contained in one program line, separated by commas. The key words of the language are written using lower-case characters, whilst variables are denoted with upper-case characters. The features of the HPL programming language are highlighted by the following program that can be used for calculating the roots of a quadratic equation:

```
 0: enp "value of A", A,"value of B",B"value of
    C",C
 1: if 4AC>BB;gto"imag"
 2: prt "real roots";spc 1
 3: prt (−B+√(BB−4AC))/2A;spc 1
 4: prt (−B−√(BB−4AC))/2A;spc 4;jmp−4
 5: "imag"
 6: prt "complex roots",spc 1
 7: prt "real","imaginary";spc 2
 8: prt−B/2A,√(4AC−BB)/2A;spc 1
 9: prt−B/2A,−√(4AC−BB)/2A;spc 4;gto 0
10: end
```

The first program line provides the input of the A, B and C variables. When the program is run, the request for the first variable's value is executed by the 'value A' message. After the answer has been keyed in, the program continues with the input of the B and C variables. In lines 3 and 4 the solution formula is given in usual form, without a multiplication sign (*). Lines 1, 4 and 9 contain branching statements together with necessary instructions.

It is apparent from the short example above that the keywords of HPL – unlike BASIC – are not English words, but abbreviations of them. This will make the HPL language somewhat artificial, but this is offset by the more economic use of the memory and by the faster operation.

3.4.4 Interfacing of the calculator/IEC system

Programmable desk calculators are now frequently used for controlling automatic measuring systems. There are obvious reasons for the rapid expansion of calculators in this application. These devices can be handled easily, can be programmed directly from the keyboard using an easily-mastered high-level language, and their capacities approach that of minicomputers.

The capacity of a calculator used in the IEC system is influenced by two important factors: the structure of the interface unit and the refinement of the firmware. In the following section the interfacing of the calculator/IEC bus will be examined with reference to these two factors, using a practical example, the Type 9825A Hewlett-Packard.

Figure 3.45 shows the block diagram of the IEC interface unit of the 9825A. This unit is located on a small printed circuit card and provides a complete mechanical, electronic and functional connection between the calculator and the IEC bus when it is plugged into one of the three I/O ports at the back of the calculator. The interface unit is of software type, the interface functions being generated by the firmware algorithms stored in the ROM controlling the microprocessor. The data and message traffic between the IEC bus and the calculator is also controlled by the microprocessor, via the I/O registers.

The IEC bus or calculator originated messages requiring interaction are monitored by separate units. The central unit of the calculator can initiate the I/O operations by addressing the interface via the command register. The microprocessor monitors the IEC bus control signals on the input multiplexer by periodic sampling. If a message requiring intervention (e.g. service request) arrives from the bus, then the microprocessor will generate the appropriate commands for transmission to the IEC bus via its receiver and driver circuits.

The possibility of external interrupt is particularly useful in the IEC system control. Figure 3.46 shows the operational speed of various devices and the I/O methods available for servicing. The programmed I/O can be used most efficiently for handling asynchronous peripherals and slow operation instruments, as the calculator can deal with other internal routines in the time interval between external device messages.

From the user's point of view one of the most important characteristics of a calculator is the efficiency of programming language control commands; these are determined by the 'intelligence' of

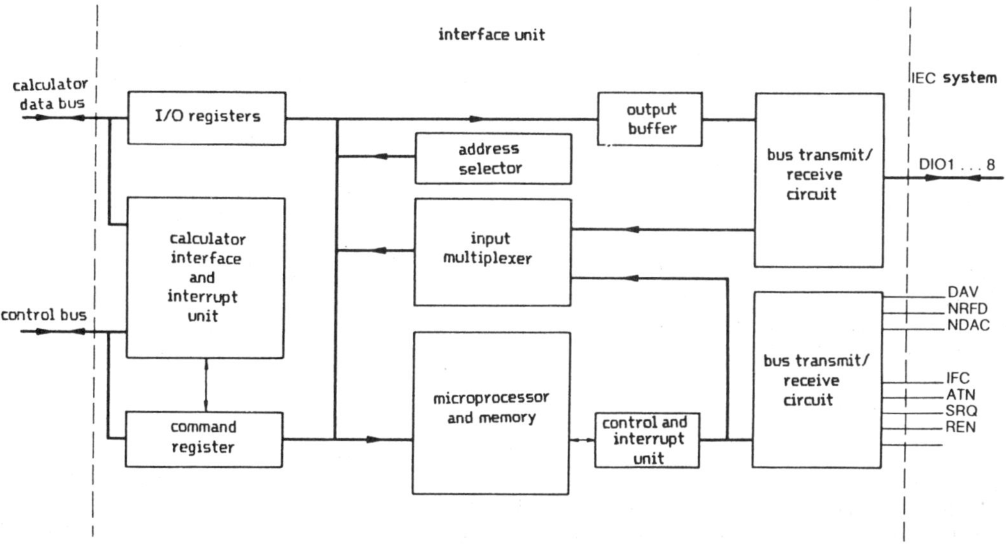

Figure 3.45 Structure of the IEC interface of the Hewlett-Packard Type 9825A calculator

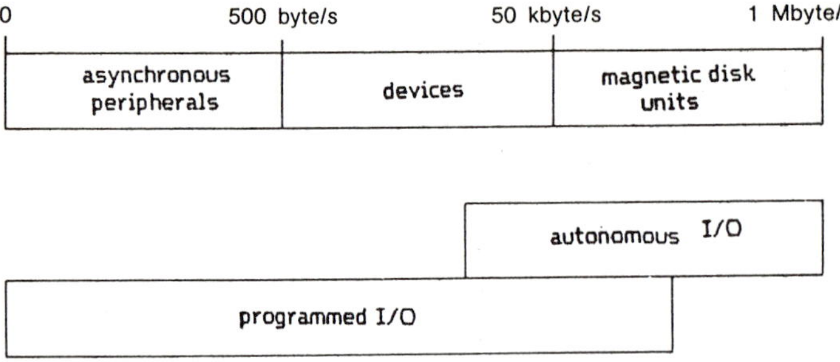

Figure 3.46 Operational speed and suitable I/O interfacing of IEC devices

the firmware. Let us examine in a few practical examples, the way efficient firmware helps the programming. The examples relate to the HP 9825A control unit equipped with I/O operation control firmware (Extended I/O ROM unit).

One of the most frequently used operations in IEC system programming is device addressing. This operation will be much simpler if the device addressing can be carried out by using a name directly. To do this, the address and the corresponding name must be identified with a so-called assignment statement (dev). For example:

dev "VOLTMETER",703,"printer",701

Similarly, the program data of the individual devices can be used in an abbreviated form in the program. For example, if a F0R4T1M3E character string is needed for the programming of a digital multimeter, the following assignment can be used:

equ "MEASLIM",F0R4T1M3E

Then the programming of the multimeter measurement limits given by the above character string can be done in the following way:

cmd "VOLTMETER","MEASLIM"

These character string assignments speed up the programmer's task and make it much easier, especially in systems using many devices, where the measuring limits are to be changed frequently.

3.5 Special IEC control computers

The main disadvantage of universal computers and calculators programmable in BASIC or other high-level languages is that they do not contain special instrument control commands in their instruction sets. Hence, in automatic measuring systems the instrument programming can be done only in an in visually, using mathematical instructions. Thus, visually the program is awkward and when modifications are necessary the identification of instruction groups is difficult.

Recognizing this shortcoming some firms already manufacture IEC system control devices equipped with task-orientated software. The most important data of a few, better-known types are listed in Table 3.2.

The capacity of the main memory can vary according to the user's needs in all of the devices and the minimum is usually 8 kwords. Without exception the devices contain some form of magnetic background storage for saving permanent data and program sections. Program editing and correction is carried out in an interactive manner using the built-in alphanumeric display and the keyboard. The devices contain a V24/RS-232-C serial interface unit in addition to the IEC interface.

3.5.1 Programming languages

The programming languages available for each machine are also listed in Table 3.2. BASIC is the most significant of these; variations of it, extended with special instrument control instructions make programming easier and the finished programs are easier to follow visually. To illustrate this point a sample program is shown, written in the special BASIC language of the Type 3530 Systron-Donner IEC controller (Figure 3.47). Table 3.3 lists the IEC-orientated instructions of the language. The following sample program is for controlling a Systron-Donner Type 7115 digital multimeter:

BUSCLR By general clearing of the IEC system, all devices are brought to ground state.

BUSREM This instruction switches all system devices to remote control.

BUSOUT!,YSCEG,0 Output of the multimeter's listener address (!) and the transfer of the program string. The meaning of the characters in the YSCEG string is as follows:

Y–DC voltage measurement,
S–10 V measurement limit,
C–setting of voltage ratio,
E–100 measurement cycles/second,
G–start.

The 0 is the delimiter of the character string.

Table 3.2 Special IEC controllers

Manufacturer and type	*Unified technologies S/4880*	*Philips PM 4410*	*Siemens S 2313*	*Systron-Donner 3530*	*Datatech SE 2650*	*Fluke 1720 A*
Main memory capacity, words	max 48K	max 64K	28K (16-bit word)	max 56K	max 128K	max 256K
Storage type and capacity, words	floppy disk 512K	floppy disk 80K	floppy disk 50K	magnetic casette 100K	floppy disk 320K	floppy disk 400K
Display	24 rows * 80 char	24 rows * 80 char	6 rows * 40 char	24 rows * 80 char	24 rows * 80 char	16 rows * 80 char
Printer	–	–	40 char	–	–	–
Interface	serial, IEC	serial, IEC	serial, IEC	serial, IEC	serial, IEC	serial, IEC
Programming language(s)	BASIC, FORTRAN IV	BASIC	BASIC, ATLAS	BASIC, ASSEMBLER	BASIC, PASCAL, FORTH, FORTRAN	BASIC, FORTRAN IV

Figure 3.47 A special IEC controller (Systron-Donner, Type 3530)

BUSIN A,D$,16 Output of the multimeter's talker address (A), and request for the 16-byte measurement datum to be put into the D string variable assigned memory field.

PRINT D$ Writing of the D string on the display unit.

Of the special controller programming languages ATLAS (Abbreviated Test Language for Avionics Systems) deserves a special mention. This program language – as its name suggests – was developed originally for automatic testing measurement of aviation equipment. Now, ATLAS is an American standard (IEEE Std 416–1976). This measuring system independent language uses English abbreviated mnemonics, for example 'ERR LMT' (error limit). Of the 238 ATLAS language elements there are 42 verbs, 35 nouns, 80 adjectives, 14 conjunctions and 67 instruction descriptions. The standard does not state the binary equivalents of the used characters. The ATLAS instructions are made up from fixed- and arbitrary-length fields. All instructions begin with a single-character flag field, followed by a six-character identification field and an arbitary-length verb field.

According to the implementation level of the language, there are three variants of ATLAS. The so-called 'Standard ATLAS' means the entire set, while the 'Subset ATLAS' contains certain sections of it in an unchanged form. The 'Adapted ATLAS' is a variant modified according to the requirements of the user.

The PASCAL and FORTH languages are suitable for real-time control and signal processing. Both are relatively easy to learn and powerful. FORTH, which is a stack-based language, is extremely fast.

3.5.2 Structure

The special IEC control units listed in Table 3.2 are essentially microcomputers built around microprocessors available on the market. The Type S/4880 control unit of Unified Technologies is outstanding, as far as structure is concerned. The block diagram of this microprocessor-equipped control unit is shown in Figure 3.48. The central control unit is a Zilog Type Z80 microprocessor, that is basically a software-compatible advanced version of the Intel Type 8080A. This processor provides the handling of the operational system and the programs used, as well as the controlling of the I/O operations.

The S/4880 contains two Type 8080A microprocessors in addition to the Z80 central processor.

Table 3.3 The special system control instructions of the Systron-Donner Type 3530 control unit

Function	*Instruction*	*Description*
Interface clear	BUSCLR	Transmission of the IFC message to the bus
Remote control	BUSREM	Switches the system devices to remote control
Local control	BUSLOC	Switches the system devices to local control
Talk	BUSOUT A$, B$, 1	Transfer of 'B' data string to 'A' address listener, with EOI message
	BUSOUT A$, B$, 0	As above, except there is no EOI message
Listen	BUSIN C$, D$, T$	Transfer of databytes from the 'C' talker address device, then input of that into the 'D' memory field to the 'T' delimiter
	BUSIN C$, D$, 14	Transfer of 14 bytes from the 'C' talker address device, then input of that into the 'D' memory field
Service request	BUSREQ X	'X' state byte
	BUSWAIT X	The assignment of an 'X' msecond time limit for serial poll
Parallel poll	BUSPOL	Start of parallel poll

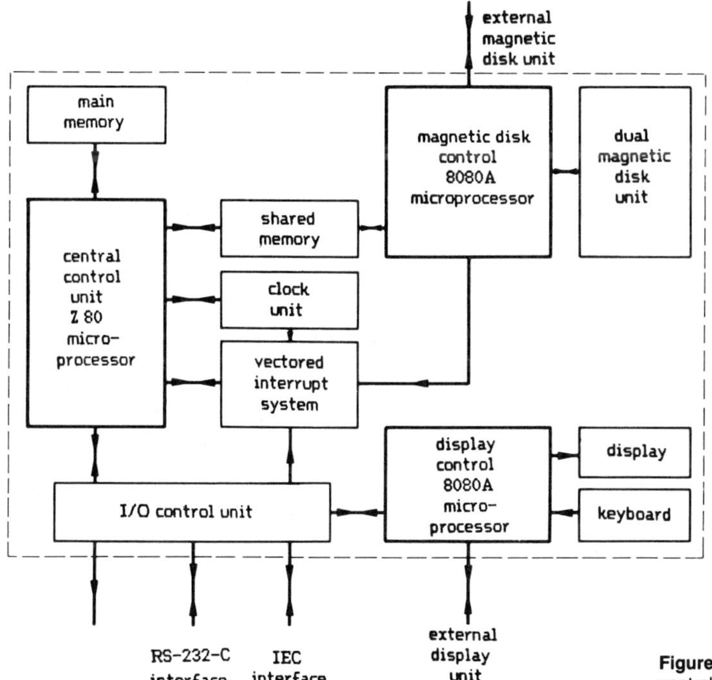

Figure 3.48 Structure of the Type S/4880 IEC controller made by Unified Technologies

One of these 8080As controls the built-in display, whilst the other handles the interfacing of the built-in dual floppy-disk system. This microprocessor equipped system has several advantages over the traditional structure. The most important is that each processor can execute certain operations independently. For example, the disk can copy a file from the main memory, while the user is keying in the next command from the keyboard.

3.6 Sequential controllers

The simplest control units of automatic measuring systems are the sequential controllers. These devices are capable of reading a previously written measurement program stored in a data memory unit. The program contains the program data of the controlled instruments and possibly some simple timing and jump instructions in a fixed sequence.

According to the method used for storing the program, there are sequential controllers containing punched card, punched tape, magnetic tape and semiconductor memory. The semiconductor memory controllers – which do not contain mechanical parts – are becoming the most widespread. In Figure 3.49 such a device, the Wavetek Type 583 Autoprogrammer Controller, is shown as a typical example.

The operation of the device can be understood by examining the functions of the keys:

- START key, begins the program stored in the memory from step 1;
- TEST key, when it is pressed followed by a two-digit (00...99) program step number, the program stops at the selected step and the program step appears on the display;
- CONT key, starts the program from the step shown on the display;
- EXIT key, interrupts program loops; after its use the program continues with the next command.

The following program segment is for controlling a function generator during a multiple-step measurement:

1 Test 01	9 Wait
2 Unit 1	10 Wait
3 Freq = 1000	11 Ampl = 4,32
4 Ampl = 5	12 Wait
5 Ofst = 0	13 Ampl = 4
6 Funct = 1	14 Stop
7 Wait	15 Go to 60
8 Wait	

The first part of the program (1...6) contains program data necessary for the first measurement step. Lines 7...10 set the time required for the execution of the first measurement. The device reads the instructions from the memory using a constant-repetition frequency (3.8 kHz). This is the reason for the inclusion of the wait instructions when a longer time is required between two steps. Each wait instruction means a 500-msecond interval.

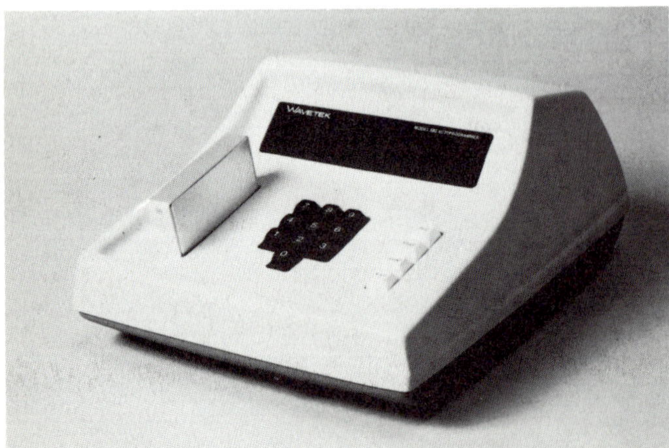

Figure 3.49 Wavetek Type 583 sequential controller

The exchangeable storage unit of the device – seen in the front – contains two Type 5204 EPROM integrated circuits (made by National). These provide a combined 1-kbyte storage capacity, allowing the storing of 100 program steps. The program input can be carried out using any EPROM programmer.

In the program steps 11...13 the decrement of the generator's output voltage occurs. As a result of the stop instruction, the control is interrupted in program step 14 and the execution of the program continues only after the CONT key has been pressed. In step 15 a jump instruction transfers the program execution into a measuring cycle beginning with program step 60.

Sequential controllers can be used most effectively for tasks involving long, unchanging measurement cycles. The modification of pre-written programs is difficult and costly. They are used mainly for controlling signal sources (generators and programmable power supply units) in repetitive, automated testing.

4 Measuring instruments, signal sources and other system elements

Around the time that the standard IEC system first appeared, another important development in measuring technology occurred, when microprocessor-controlled 'intelligent' instruments began to become available. Instruments with built-in microprocessors were first used in analytical chemistry applications. Formerly, in chromatography and photometric measurements which require many arithmetic operations, calculators or minicomputers were used to separately evaluate the results. The integration of small cheap microprocessors into the instruments made these separately-purchased expensive devices redundant.

Furthermore, the microprocessor could be used for controlling the instrument operation, as well as simultaneously carrying out the calculation tasks. Thus the instrument's face changed, there was a decrease in the number of handling devices and the display contained textual messages in addition to numeric data.

With the combination of the two different functions, the control and the calculation, the on-line microprocessor could modify its own operation without the intervention of the user by making decisions according to the intermediate measurement results. Thus the execution of a complicated analytical task, consisting of several hundred steps, could be carried out by pressing a single START button, following which the instrument then automatically carried out all operations, including printing out the results.

The application of microprocessors in more general electronic instruments began soon after this. Owing to the system characteristics of electronic measurements, the use of microcomputers is advantageous for yet another reason; they can also be used for controlling the interface that connects the instruments.

In the following section microprocessor controlled instruments will be discussed generally, first by examining the advantages of programmed control and the new possibilities created by it.

Then a few microprocessor-controlled instruments will be described in detail.

Finally, the modern hardware elements that facilitate the microprocessor-controlled instrument/IEC bus interfacing will be dealt with.

4.1 Microprocessor-controlled instruments

The practical application of digital circuits in measuring technology began in 1952. It was then that the first custom-designed digital voltmeter made by the American Non-Linear Systems company became available. The use of digital display spread quickly, as it had several advantages over the traditional analogue display. The resolution achieved was much greater and the constraints on measurements were reduced significantly. Furthermore, the method completely eliminated the subjective measurement reading errors associated with analogue techniques. Another advantage was that the measurement data, converted to binary coded format for the numeric display, could be directly output from the display-decoder unit at the back of the instrument; thus the measurement result was also available in the form of digital signals.

A general instrument system was developed that had the following basic features. The electrical signal proportional to the measured quantity was converted into the digital signal format necessary for the display after analogue processing (Figure 4.1(a)). Another important characteristic of this system was that the intervention elements (e.g. switches, potentiometers, condensers, etc) could be operated directly, using the handling devices located on the instrument faces.

From the mid 1950s digital circuit techniques and technology have been developing rapidly, with a consequent improvement in the properties of digital instruments. However, only the appearance and incorporation of microprocessors brought about a fundamental and structural change by allowing operation to be controlled by a program.

The measuring technology application of microprocessors and microcomputers is not limited to the

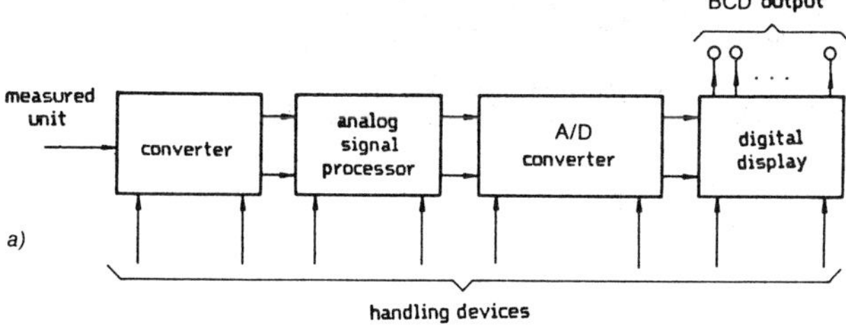

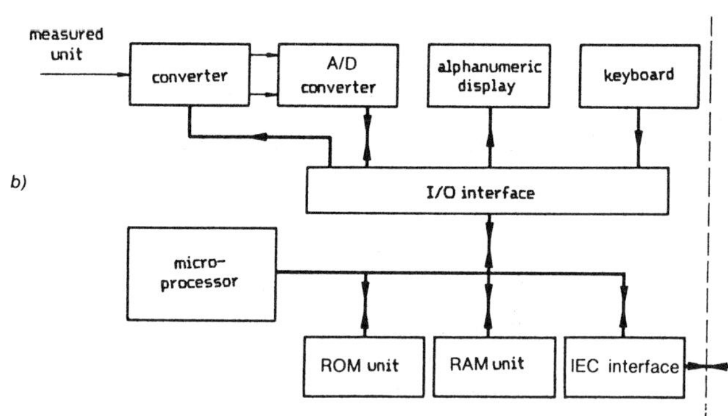

Figure 4.1 Structure of digital instruments. (a) Traditional structure, (b) programmed control

execution of arithmetic tasks only. These units can be used successfully as controllers, in place of the logic circuits consisting of traditional gates and flip-flops. The basis of programmed control is that the signals necessary for the required control tasks can be generated from the appropriate contents of ROM connected to the microprocessor.

Table 4.1 contains approximate comparisons between the ROM capacity and the number of gate circuits that can be replaced by it. It can be seen that a 32-kbit-capacity ROM can simulate the operation of between 2048 and 4096 gate circuits. Considering that nowadays this ROM capacity can be put into one or two integrated circuit chips, and bearing in mind that 200 to 400 SSI integrated circuits would be necessary for the control task executed by it, the magnitude of development brought about by the introduction of programmed control in measuring technology can be readily appreciated.

Table 4.1 Substitution of wired control with programmed control

ROM capacity, bit	*Number of equivalent gate circuits*	*Number of equivalent IC chips*
2048	128 . . . 256	13 . . . 25
4096	256 . . . 512	25 . . . 50
8192	512 . . . 1024	50 . . . 100
16384	1024 . . . 2048	100 . . . 200
32768	2048 . . . 4096	200 . . . 400

Figure 4.1(b) shows a typical block diagram of microprocessor-controlled instruments. The biggest difference from the traditional structure digital instrument is that the A/D conversion is executed at the input of the instrument; thus all further processing is executed in digital form. The handling devices found on the face of the instrument are not used for intervening with the operation of the circuits but for instructing the microprocessor. Similarly, the display unit can show not only the numeric value of the measured quantity, but various alphanumeric messages as well.

4.1.1 The advantages of programmed control

Let us examine in detail the advantages of the application of programmed control in the instruments.

(1) *New measuring possibilities:* The microprocessor built into the instrument can control measurement operations (e.g. tuning, measurement limit change, calibration, etc), frequently-used measurement routines (e.g. signal-noise ratio measurement, etc) and simple calculations automatically. The latter also provides an opportunity for the automatic measurement or calculation of specific quantities, such as bandwidth.

The measurement results can be stored in the instrument's memory and thus the instrument operator can measure or average characteristic values (maximum-minimum values, for example) without intervention during the chosen time interval.

(2) *Simpler handling:* One of the biggest advantages of programmed control is that the most convenient form of display and choice of handling devices can be used in the design and construction of the instrument. This is due to the ability of the control program to separate the handling (operational) requirements and the intervening hardware elements completely. Thus the operation of a single key can start a number of independent operations. The program control can also, for instance, change the operation of the instrument between operator intervention, according to the measured values. As a result of these features, the number of handling devices is reduced, whilst their functions will be related to the measuring task. In most microprocessor-controlled instruments the testing of handling device settings is automatic and a special display warning indicates meaningless or contradictory settings.

Automatic calibration is one of the operational simplifying possibilities that deserves special attention. This means that at certain intervals (e.g. at every switch on) the microprocessor checks the operation of units influencing the measurement error, without the intervention of the operator, and corrects small deviations. The block diagrams shown in Figure 4.2 illustrate the automatic calibration process. Figure 4.2(a) shows the measurement configuration and Figure 4.2(b) the calibration arrangement. During calibration the microprocessor switches the reference to the input unit of the instrument, simultaneously interrupting the connections between itself, the handling devices and the display. The validation occurs according to the values stored in the memory. The instrument uses some form of error message to indicate that the prescribed tolerances have been exceeded. Drift occurring in the instrument can be corrected by the same process when the value of the reference is zero.

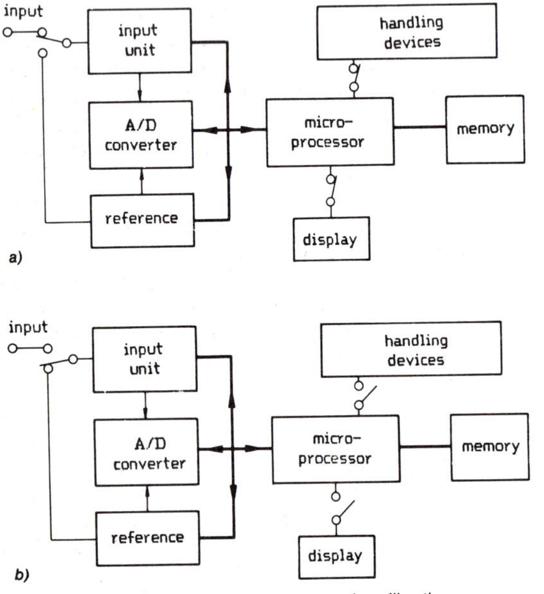

Figure 4.2 Operational phases of automatic calibration instruments. (a) Measurement, (b) calibration

(3) *The IEC interfacing becomes simpler:* An increasing number of semiconductor manufacturers recommend a special IEC interface unit for their microprocessors. Apart from some auxiliary functions, these LSI circuits contain the complete IEC bus/instrument interface. These circuits will be discussed in detail later.

(4) *More reliable operation:* The main source of errors in electronic devices is the unit–circuit connection. In microprocessors and other LSI units the internal connection, the production of the metal layer on the silicon wafer is carried out under strictly-controlled manufacturing conditions. It is due to this and the rigorous testing following the manufacture that the devices built from LSI circuits are considerably more reliable than traditional SSI- and MSI-based devices.

(5) *Smaller size and weight:* Figure 4.3 shows the internal construction of a traditional impulse generator. The digital control requires several hundred integrated circuits, the supply source is heavy and very large owing to its high power consumption, occupying about one third of the instrument's internal volume. The smaller size and weight of programmed control instruments is due to two factors; the first factor is that the digital control circuits occupy less space, and the second is the significant reduction in the size and weight of the supply unit due to the smaller power consumption of the circuits used.

(6) *Favourable price* (for appropriate scale production): The microprocessor is essentially a universal control unit that raises the accomplishment of the individual special requirements from the hardware level to the software level. The same microprocessor can handle entirely different tasks using different software.

The hardware costs are only part of the manufacturing costs of programmed-control

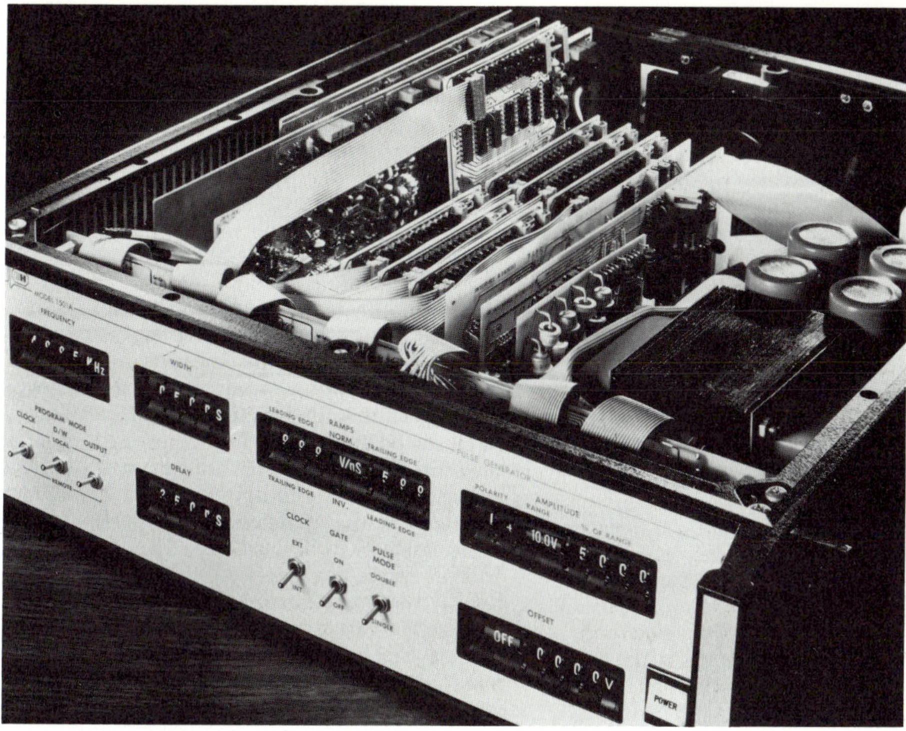

Figure 4.3 Internal structure of a traditional impulse generator (E-H Research, Type 1501A)

digital devices. The design and testing of ROM-stored firmware is very laborious and costly for the manufacturer; thus, programmed control is economical only for reasonably large-scale production.

4.1.2 The disadvantages of programmed control

(1) *It is slower than traditional control:* The operation of microprocessors is sequential, i.e. they can execute only one task at a time. A task series can consist of requests for data from an I/O gate, testing of the incoming data, output of the decision made, the response or the transmission of the data to another I/O gate. As opposed to this, the wired logic control generally operates in parallel mode and is therefore very suitable for executing multiple tasks simultaneously.

Apart from the inherent slowness, the technology used in circuit manufacture also plays an important role. Usually, fast bipolar circuits (TTL, ECL) are used for traditional control, while the cheap, few-component microprocessors are made using an MOS technology.

Usually the relative slowness of programmed control is not noticed by the user. The various phases of instrument operation are determined either by the temporal change of the physical quantities to be measured or by the intervention of the operator; therefore the speed of the microprocessor usually does not limit the effective operational speed of the instrument.

(2) *Repair of faulty instruments is more difficult.* The major disadvantage of microprocessor control is that the testing of the instruments is very difficult during manufacture or later. One of the reasons for this is that there are differences between a ROM-simulated special circuit and its traditional equivalent, built from gates and flip flops. A more important difference is that the traditional circuit element, an AND gate for example, is always present in the circuit, while the ROM-generated AND function will perform its duties only when the program control assigns it. Another problem arises from the bus system structure of the microprogrammed devices; in most cases it is difficult to ascertain the identity of the unit transmitting erroneous signals of the bus. It is also frequently very difficult to distinguish

between hardware and software errors, since such errors can also be data- or time-dependent.

The functional testing of such systems involves the checking of long signal sequences, frequently consisting each of several thousand bits. The problem is aggravated by the fact that in most cases these signals are not repetitive.

Recognizing the problems arising with the repair of microprocessor-controlled instruments, the manufacturers try to incorporate self-checking, auto-check routines in the program. Self-testing applies at different levels and in some instruments, despite special test routines, only the fact that there is an error occurring is displayed. However, it is possible to develop error-identifying routines that find the faulty functional element or printed circuit card.

4.2 Instruments and signal sources

The wide variety and multitude of electronic measurement tasks does not permit us to give a comprehensive description of instruments and signal sources used in automatic measuring systems. Accordingly, we have chosen to discuss the present trends and results of instrument development through the detailed description of two modern programmed-control instruments. In the course of the discussion we emphasize the special features provided for the user and the way these features are achieved by internal system organization and design.

4.2.1 Digital voltmeter (Solartron Type 7065)

The Solartron Type 7065 digital voltmeter (Figure 4.4) is an instrument suitable for measuring direct or alternating voltage and resistance; its resolution is 6½ digits. Apart from straightforward measurements, eight different measurement and calculation routines can be programmed either using the 18 keys located on the instrument face, or via the IEC interface, utilizing the arithmetic capability of the microprocessor that handles the internal control of the instrument.

The programs are compiled in Table 4.2 where symbol R denotes the result of the operation, while x denotes the variable. The programs contained in the Table can be made into a chain. In the first segment of the program chain, x is the instrument's measurement data, while in the following program segment the variable is the first result.

Figure 4.4 Microprocessor-controlled Solartron voltmeter, Type 7065

Table 4.2 Measurement programs for the Solartron voltmeter, Type 7065

Serial number	*Program name*	*Variation*	*Operation*	*Execution time msecond*
1	multiplication	–	$R = cx$ where c = constant	12
2	percentage difference	–	$R = 100\,\frac{x-n}{n}$ where n = nominal value	14
3	offset	–	$R = x - \Delta$ where Δ = offset	2
4	ratio	0	$R = \frac{x}{r}$ where r = reference value	12
		1	$R = 20\lg\frac{x}{r}$	135
		2	$R = \frac{x^2}{r}$ where r = resistance value	20
5	maximum/minimum	0	$R = x$	6
		1	$R = \max x$	6
		2	$R = \min x$	6
		3	$R = \max x - \min x$	9
6	limits	–	see text	10
7	statistics	0	$R = x$	24
		1	$R = \frac{1}{i}\sum_{k=1}^{i} x_k = \bar{x}$	40
		2	$R = \frac{1}{i}\sum_{k=1}^{i} (x_k - \bar{x})^2$	100
		3	$R = \sqrt{\left[\frac{1}{i}\sum_{k=1}^{i} (x_k - \bar{x})^2\right]}$	190
		4	$R = \sqrt{\left[\frac{1}{i}\sum_{k=1}^{i} x_k^2\right]}$	210
8	thermocouple	0	Cu/Co	70
		1	Pt/PtRh	70
		2	Fe/Co	70
		3	NiCr/NiAl	70

Some of the programs have several variants and the user must decide which one is to be used. The instrument programming is executed by using the numeric symbols listed in the Table in the following way:

PROGRAM
4
ENTER
1
ENTER
17.32
ENTER

The above program is used for calculating voltage ratios expressed in decibel units (program 4, variant 1); the reference value is 17.32 V.

Of the programs listed in the Table, the No 6 'limits program' requires some explanation. In this program, after upper and lower limits have been assigned, the instrument counts the measurement results relative to the limits (Figure 4.5). After the program has been run, the following data can be called on the instrument's display:

- the number of measurements above the upper limit,
- the number of measurements below the lower limit,
- the number of measurements between the two limits,
- the value of the upper limit, and
- the value of the lower limit.

The data are displayed in sequence on the repeated operation of the RECALL key.

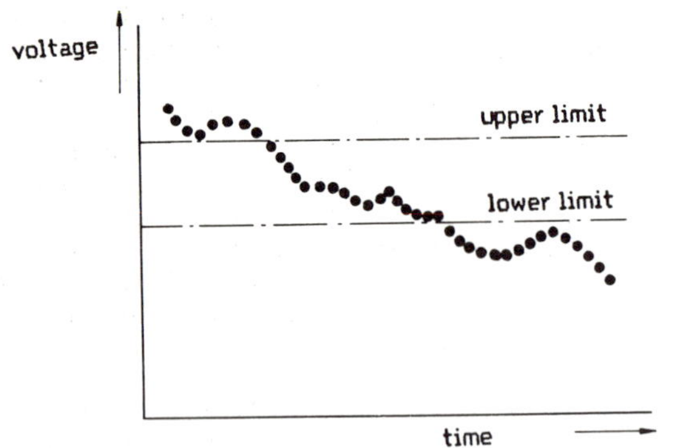

Figure 4.5 Measurement of the voltage by comparison to the two assigned values in the No. 6 delimiter program by a Solartron voltmeter, Type 7065

4.2.1.1 *Structure and operation of the instrument*

The block diagram of the 7065 is shown in Figure 4.6. The instrument can be divided into two functional parts; this is denoted with a broken line in the Figure. The upper part contains the analogue unit electrically connected to the unit under test and the logic that controls its operation directly. The input unit contains the input protection circuits, the operation selector relays and their control logic. The current generator necessary for the resistance measurements and the converter used during alternating voltage measurements are also directly connected to this unit. The task of these two devices is to generate a direct voltage proportional to the measured unit which is then transferred to the amplifier and measurement limit changer. This unit provides the appropriate level of direct voltage for the A/D converter.

The amplifier and measurement limit changer unit contains an automatic zeroing circuit for the compensation of drift in the input amplifiers and the integrator during direct voltage measurement. The automatic zeroing cycle is commenced by the following events:

- measuring limit change,
- integration time change,
- operational mode change, and
- after 10 seconds from the previous zeroing cycle have elapsed.

The A/D conversion method used in the Type 7065 is a modified version of traditional dual-slope integration. The method is based upon the generator impulses which have a duration proportional to the direct voltage presented to the converter's input, and these gate the internal clock signal counter. The integration period of the measurement is chosen by

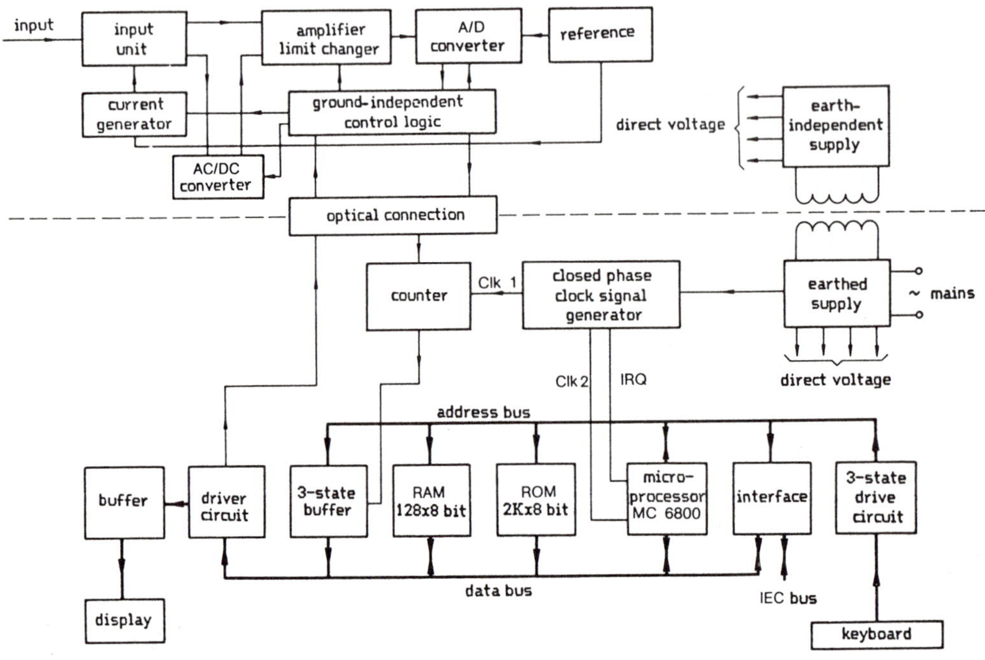

Figure 4.6 Structure of the Solartron Type 7065 voltmeter

the user according to the resolution required. The resolution is 3 digits for 1 msecond, which is a short integration period. Increasing the integration time improves the resolution of the instrument and 6 digits can be achieved using a 1-second integration period.

An interesting feature of the solution is that the clock signal to be counted (clk 1) is forced to a value that is a multiple of the power supply frequency by a closed-phase loop. The advantages of this method will be discussed in detail later under the heading of noise rejection.

Of the other units of the isolated (ground-independent) section of the instrument the unit generating the reference signal influences the measurement error of the instrument greatly. This direct voltage must be highly accurate and stable, as it is required for the operation of the A/D converter and the resistance converter.

The analogue section's operational control logic has three important tasks:

- The decoding of operational mode and measurement limit information arriving from the digital control unit and the control of the intervening devices (signal receivers, FET switches, etc).
- The transmission of the counter gating signal, generated by the A/D converter, to the counter.
- The optical and transformer isolation of the grounded and ground-independent units of the instrument (the manner in which this separation is achieved will be discussed later in the section dealing with measurement methods).

The digital control unit of the Type 7065 is bus based, using programmed control. The microprocessor is the Motorola Type 6800 which is popular as an instrument controller due to its comparatively simple interface requirements and programming. The task of the microprocessor is to control the operation of the instrument according to the instructions arriving from the keyboard or from the IEC interface and the data stored in the ROM.

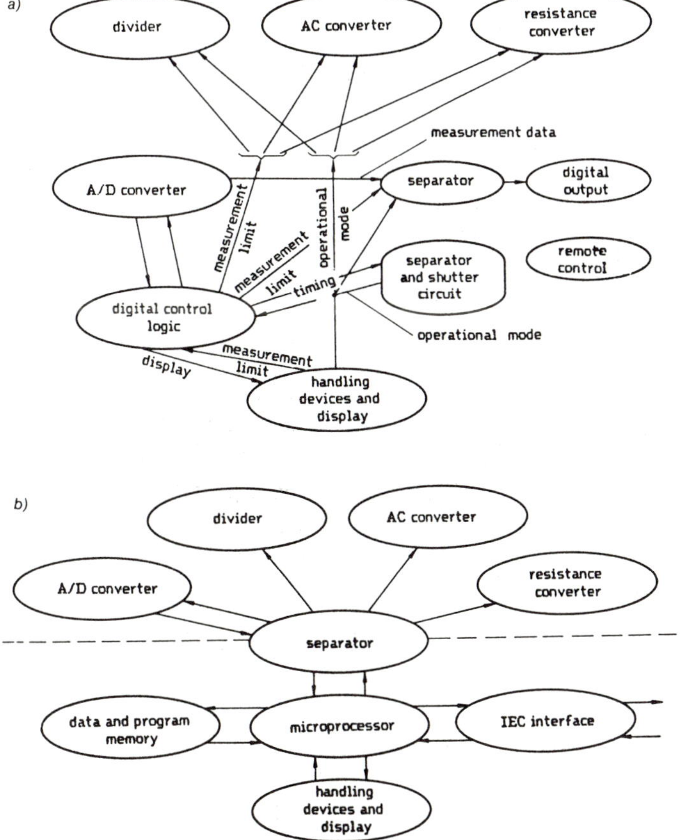

Figure 4.7 Functional structure of digital voltmeters. (a) Traditional organizing, (b) programmed control

Apart from the microprocessor and the ROM, the measurement-data-storing RAM and the buffers and drive circuits interfacing the display and the analogue units are also directly connected to the data and address buses.

Another important unit of the digital control system is the closed-phase clock generator. Its task, apart from generating the clock signal required for the operation of the A/D converter, is the generation of the clock signal controlling the microprocessor and the generation of the interrrupt (IRQ) signal. The latter interrupts the operation of the microprocessor every 1.25 mseconds. The reception of new information arriving from the keyboard of the IEC interface and some other house-keeping operations take place in these 'intervals'.

The bus system control applied in the Type 7065 simplifies the structure of the instrument significantly. In traditional wired-control voltmeters the signal traffic between the functional units takes place on a complex cable system (Figure 4.7(a)). In the 7065 and in other programmed-control instruments, however, the individual units are connected to the microprocessor in a star configuration. This configuration is very advantageous in two respects; first, it is easy to interface the individual units; secondy, the necessary separation in 'noisy' measurements can be achieved with minimal effort (Figure 4.7(b)).

The microprocessor built into the instrument is used for executing the error-search routines and testing the instrument for correct operation, as well as for the controlling of the measurement cycles and evaluation programs. On keying in the test program given in the instruction manual of the instrument, the displayed error messages give the sources of possible errors broken down to functional units, or even to component level in the digital control unit.

4.2.1.2 *Interface units*

The 7065 has three different interface options:

- BCD parallel interface
- RS-232-C serial interface, and
- IEC interface.

The RS-232-C and the IEC interfaces can also be used together. In such cases the instrument can be controlled from the RS-232-C interface as well as transmitting data on it. However, the IEC interface has priority in control and messages arriving on the IEC interface therefore forbid the reception of further RS-232-C messages.

The IEC interface unit of the instrument realizes the following functions:

SH1 source handshake
AH1 receptor handshake
T5 talker
L3 listener
SR1 service request
RL1 remote/local control
PP1 parallel poll
DC1 device clear
DT1 device start.

Using the above functions the Type 7065 can operate as an IEC device in two operational modes. It can be connected to any IEC-compatible data collector if the two-way switch, located on the device, is in the 'talk only' position. A simple, controller-less system allows the automatic registration of the measurement data of the instrument.

Alternatively, in an IEC system containing a separate controller, the Type 7065 can also be automatically programmed for measurements and calculations, in addition to the output of data in the desired format.

Table 4.3 Code table of the measurement programs of the Type 7056 voltmeter

Alpha character	*Possible values of n*	*Name of function*	*Variants*
B	0,1	Data format	0 = decimal, 1 = binary
D	0 . . . 3	Display format	0 = 3 digit, 1 = 4 digit, 2 = 5 digit, 3 = 6 digit
F	0,1	Screening	0 = yes, 1 = no
H	0,1	Handshake	0 = continuous measurement, 1 = measuring cycles are controlled by the handshake process
M	0 . . . 2	Operational mode	0 = DC voltage, 1 = resistance, 2 = AC voltage
R	0 . . . 7	Measurement limits	0 = autorange, 1 = 10 MΩ, 2 = 1 MΩ (1000 V . . . 7 = 10 Ω) 0.01 V

Programming: All functions of the Type 7065 can be controlled via the IEC bus using an appropriately-constructed character string. The code system containing the program characters is compiled in Table 4.3. The individual functions and their variants are assigned with an alpha character and a number (n) following it. The program instruction can be 'M0R0F0D3', for instance, when the Type 7065 will measure direct voltage using automatic ranging and employing screening, and the display will be given in 6 digits. The controller must transmit the listener address of the instrument with the ATN = 1 and must then indicate that a device-dependent message transfer is to follow (ATN = 0) before the program instruction is sent. The measurement can be commenced with the GET message.

The calculation or data processing operations of the Type 7065 can be programmed similarly with the character string assigning the arithmetic operations being sent together with the character string programming the measurement.

Data transmission: The transmission of measurement data uses the format appropriate to the IEC proposals. A typical message could be the following:

$$\underbrace{\triangle}_{1}\underbrace{VDC}_{2}\underbrace{\triangle}_{3}\underbrace{+}_{4}\underbrace{1.23456E{+}03}_{5}\ \underbrace{crlf}_{6}$$

where

(1) signals the validity of measurement data (no overload),
(2) are the symbols of the operational mode,
(3) is the separating character,
(4) is the sign,
(5) is the measurement data in floating decimal point format, and
(6) are the delimiters.

Should the need arise, the above message format can be simplified by the omission of the introductory part (1, 2, 3). By doing this the time required for data transfer and the memory required for storing the data are each decreased by about a third.

4.2.1.3 Characteristics

In using a digital voltmeter or any other instrument in an automatic measuring system, we must know the factors limiting its use and the effect of its environment on the measurement accuracy. On the data sheets detailing the characteristics of the instruments there are usually no explanatory notes given regarding the characteristics quoted. In the following, while the most important characteristics of the Type 7065 are given, a general discussion of a few fundamental notions found in most digital instrument data sheets will also be included.

Resolution: The resolution of an instrument is the quotient of the smallest and largest measurement value that can be displayed within a measurement range. The resolution of digital instruments depend solely on the number of digits of the display. To reduce the number of measurement range changes, the highest magnitude (most significant) digit of the digital instrument display is called the 'half digit'. For example, on a 3½-digit display the maximum measurement value is 1999. This over-range half digit is usually ignored when the resolution is calculated. Accordingly, the resolution of a 3-digit voltmeter is 0.1%, a 4-digit number results in 0.01%, a 5-digit one in 0.001% and a 6-digit one in 0.0001% resolution.

When necessary resolution for the given measurement task is considered, it must be kept in mind that the above percentage resolutions refer to the full-scale value. For example, with a 3½-digit voltmeter measuring in the 10 V range and relating the resolution to a 4 V voltage value, the resolution is only 0.25%.

Sensitivity: The sensitivity of a digital voltmeter is the smallest voltage change the instrument is capable of detecting. In mathematical terms the sensitivity is the product of the resolution and the full-scale value of the smallest measurement range. The sensitivity of the Type 7065 is 1 μV.

Error and stability: The error of a voltmeter is the difference between the displayed and the correct voltage values, compared to the correct voltage under defined reference conditions (e.g. duration, temperature, humidity, supply voltage, etc).

During operation in environmental conditions different from the reference conditions the displayed value may keep changing. This temporal change is specified as stability by the manufacturers.

Usually two stability values are given for digital voltmeters, one is the short-term stability over 24 hours, whilst the other is the long-term stability over 90 days or perhaps six months. Solartron give the DC long-term stability value of their Type 7065 voltmeters as ± 0.004% + 4 digits for six months.

Obviously, the error of an instrument cannot be less than the error of the standard against which it was calibrated, and it will achieve its most accurate state directly after calibration. Instruments, including digital voltmeters, are calibrated in the standards laboratories of the manufacturers. Their own voltage standards are calibrated against national or international standards several times each year. Solartron, for example, calibrate their voltmeters against a 2 ppm accuracy standard at the end of their production. The accuracy of the instruments can be maintained with regular recalibration. The necessary frequency of recalibration is determined by the stability of the instrument.

Noise rejection: Noise rejection is necessary for two different types of electrical noise.

Normal or *series mode noise* is superimposed on the measured signal as presented to the input of the instrument. Separation of the noise from the measured direct voltage can be achieved either by using a low-pass filter or by integration. If the noise has a wide bandwidth, filtering is the preferred solution, but it suffers from the disadvantage that the measurement rate is slowed down. In practice, interference with the measurement frequently originates from the mains supply and its frequency is therefore a known and more or less constant value. One of the main advantages of dual integration (the A/D conversion is based upon a voltage/frequency conversion) is that very large noise suppression can be achieved if the integration time is a multiple of the cycle time of the noise signal. As the frequency of the AC mains often differs from the nominal 50 Hz value, the noise suppression can be improved greatly if the clock signal used for the integration and the mains signal have some kind of synchronous connection. In the Type 7065 this is ensured by the regulating circuit operating as a closed-phase loop. Figure 4.8 shows the noise suppression of an integrating A/D converter as a function of frequency, with and without filtering. The value of normal mode noise suppression (NMR) is strongly frequency-dependent and is therefore given either diagrammatically or relative to certain frequencies.

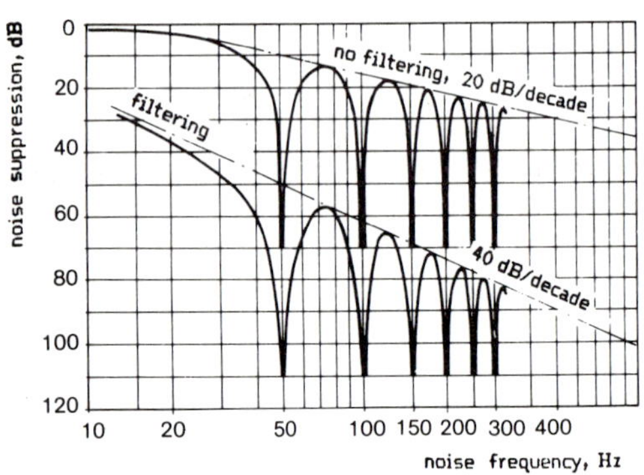

Figure 4.8 Noise suppression of the dual integration A/D converter as a function of frequency

Another type of noise signal is *common mode noise*, which appears between the input points of the instrument and the casing of the instrument. Common mode noise is characterized by having the same value on both input lines relative to earth. It is caused by the voltage difference (U_Z) between the instrument's casing and the measured circuit or device (Figure 4.9 (a)). The symmetrical common mode noise will not interfere with the measurement on its own; however, the problem is caused by the inequality of the resistance values of the two lines

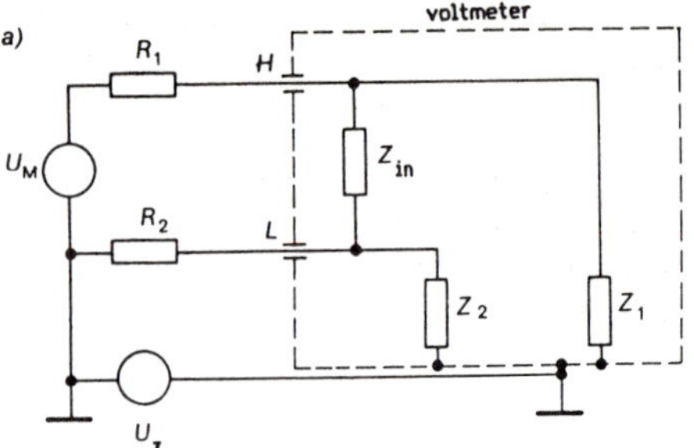

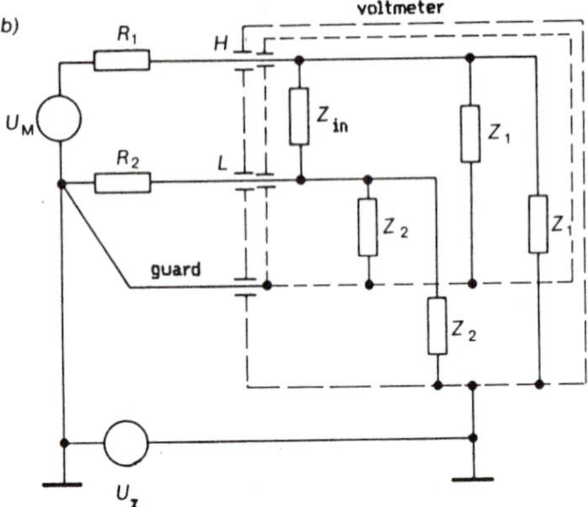

Figure 4.9 Reduction of noise sensitivity of digital voltmeters using guarding techniques. (a) Normal arrangement, (b) application of guard

within the measuring cables ($R_1 \neq R_2$), whereby the common mode current flowing through them converts the common mode noise into normal mode noise in proportion to the degree of asymmetry.

Common mode noise can be reduced very efficiently by guarding, which is used almost without exception in every digital voltmeter. The basis of the method is that the two input points of the instrument (high, H, and low, L) and the circuits connected galvanically to them are separated by shielding from the earthed case of the instrument (Figure 4.9(b)). Thus the impedance between the measuring circuit and ground is increased, because the common mode current passing through the measuring line will be determined only by the very high isolating impedance of the guard circuit.

The shielding must be provided for the entire measuring circuit and the shielding of a ground-independent part therefore has a 'guard' output or terminal on the instrument front to provide a shielding facility for the external measuring lines as well. The GUARD line must be connected to the ground of the measured source (Figure 4.9).

When the common mode noise rejection (CMR) is given, the manufacturers assume that there is a

1 kΩ (or occasionally 100 Ω) asymmetry producing resistance in series with the L input. To compare the two definition methods, the CMR value referring to 100 Ω should be reduced by 20 dB. The Type 7065 voltmeter's CMR value is not less than 150 dB for DC voltages, based on 1 kΩ asymmetry.

Measurement rate: One of the most important characteristics of voltmeters used in automatic measuring systems is the maximum reading rate (maximum number of measurements taken in a second). Let us examine the factors influencing the operational speed of digital voltmeters in direct-voltage measurements. In listing the individual components, the data relevant to the Type 7065 will be given, as a matter of interest. The timing relationships are given in Figure 4.10, where

t_a is the settling time of 2.7 mseconds.

t_b is the range change delay, between 12.4 and 35.4 mseconds – this factor should be included only if there is a need for range changes, when its value will depend upon the number of steps necessary to find the appropriate range.

t_c is the integration time, between 0.3 msecond and 1.28 seconds, depending on the resolution and the noise suppression required. Better resolution and higher noise suppression can be achieved only by increasing the integrating time and by using guarding. These slow down the measurement considerably and for this reason, the user of modern voltmeters has a choice between fast operation and resolution in varying degrees. Of the data relating to the Type 7065 the 0.3 msecond value relates to 3-digit unscreened measurement, while the 1.28 seconds is the integration time of a 6-digit measurement using a filter.

t_d is the delay time of automatic calibration (calibration time), and is 1.6 mseconds longer than the integration time.

t_e is the data conversion time, between 0 and 2.5 mseconds. On the output of the A/D converter the data are available in pure binary form. If the data output is carried out using another format (BCD or ASCII), then the conversion will take some longer time.

t_f is the display update time of 4 mseconds.

Accordingly, the total time of the measuring cycle is:

$$t_t = t_a + t_b + t_c + t_d + t_e + t_f$$

4.2.2 Programmable power supply (the Kepco SN-488 system)

Several manufacturers are currently marketing programmable IEC-compatible supply sources, and the highly-modular Kepco SN-488 system (Figure 4.11) is an outstanding example of these various products. This particular device is a very good example of existing product adaptation to a new measuring principle, i.e. IEC compatibility with control features.

4.2.2.1 Structure and operation

The system of the Kepco Type SN-488 power supply unit is a universal, module-structured device family that allow the programming of 16 independent

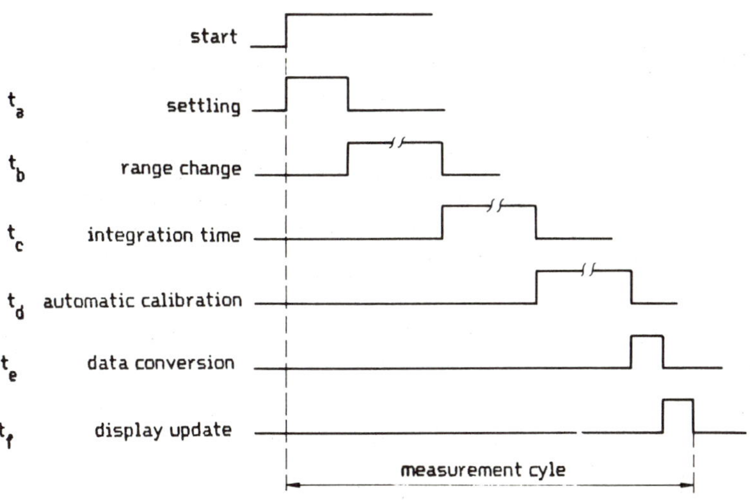

Figure 4.10 The complete measurement cycle of the Type 7065 Solartron voltmeter

Figure 4.11 IEC-compatible Type SN-488 Kepco power supply unit

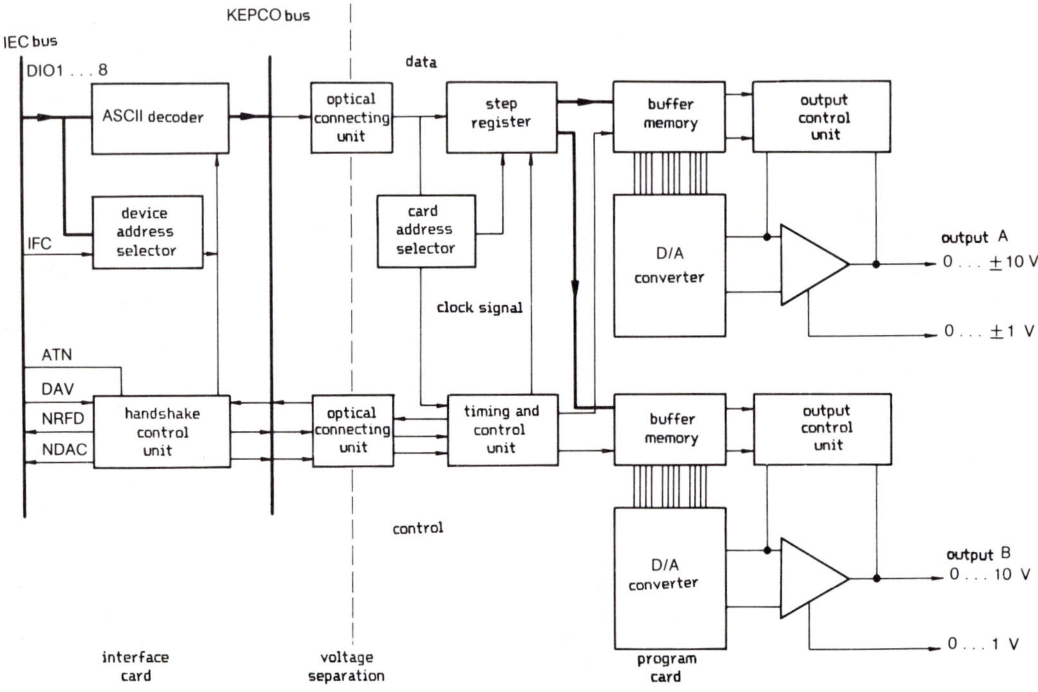

Figure 4.12 Block diagram of the Type SNR-488-4 Kepco interface

power supply units via a single IEC interface. The system is based upon a Type SNR-488-4 interface unit that can contain one IEC interface card and four SN-488-B program cards. All program cards have two independent analogue outputs that can provide 0–10 or 0–1 volts for the programming of the supply voltage. The output of the interface card, when directly connected to the IEC bus, can control a further four program cards via the internal Kepco bus.

Figure 4.12 shows the detailed block diagram of the interface unit. The interface card on the left of the diagram provides the IEC interface AH1 receiver handshake and the L1 listener functions. In addition the interface unit is also suitable for the 'listen only' operational mode by manual switching and can therefore be used in controller-less systems as well.

From the data signals arriving on the IEC bus, the allocated device address is selected by the device address selector, which will then start the ASCII decoder unit. This is basically a ROM that converts the ASCII-coded data of the IEC bus into serial, hexadecimal code. The interface card also contains the handshake control unit that controls the timing of the data transfer cycles.

The output of the interface card is connected to the input of the program cards via the internal Kepco-bus. The diagram shows the structure of one program card; in the complete system eight similar units can be controlled with a single interface card.

The voltage separation between the program card and the Kepco-bus is achieved by opto-isolators; thus, the analogue output voltage of the individual program cards can be either floating or fixed with respect to an arbitrary potential. The identifying address of the program card is recognized by the address selector unit and will start the input of the program data into the step register. The writing is controlled by the 2-μsecond time signal generated by the timing and control unit. The programming of both program card channels (A and B) is simultaneous. The program data, converted to parallel format, are transmitted from the step register to the buffer and then to the 12-bit D/A converter. The latter generates the accurate analogue voltage (0.0244% resolution) driving the output amplifier.

The polarity (+ or −) and amplitude (10 V or 1 V) of the analogue output voltage is selected by the output control unit, according to the program data. The analogue voltage outputs can be loaded up to 400 mA and the control of a power supply unit, programmable with an arbitrary voltage, is therefore possible. The A and the B outputs can also be used for arbitrary control. One possible approach is for one of the outputs to control the voltage of the supply source, whilst the other sets its current limit.

4.2.2.2 Programming via the IEC interface

The SN-488 system can be controlled by an 11-character string via the IEC bus. An example of such a character string is:

?%5=0999500

The first character of the program string, (?), is the 'unlisten' (UNL) message of the IEC interface that returns the previously listener-addressed device to the ground state. The second character, (%), is the device address of the SN-488 system, which is one of the 31 possible listener addresses. The third character, (5), is the address of a program card, which is one of eight possible addresses. The next character, (=), serves the purpose of separating the address and the data.

The first character of the data group, (0), is the so-called 'control character', determining the polarity and the amplitude of the two program-card outputs. The possible variants are compiled in Table 4.4, in which the H symbol denotes 0.... ± 10 V, and the L symbol denotes 0....± 1 V amplitude.

The control character is followed by two strings, each consisting of three characters; these set the precise voltage value of the two program card outputs. In our example, the first string (999) sets the voltage of the A output to a maximum value (10 V or 1 V), while the second string (500) generates 5 V or 0.5 V voltage on the B output, according to the value of the control character.

The above program string has to be in a strict format. If it contains more or less characters than it should, or if the separating character is omitted from between the address and the data, it will be rejected by the device.

4.2.2.3 Characteristics

In the following, which is still concerning the SN-488 system, a few important characteristics relating to the program control of supply sources will be discussed. Numeric data will not be given in the discussion, as the above-mentioned system has a modular structure and can therefore be used for the control of any supply source that has technical specifiations appropriate to the given application.

Source effect, line regulation: This is the maximum value of the output voltage change caused by an input (mains) voltage change under constant-load conditions.

Load effect, load regulation: This is the maximum value of the output voltage change caused by a load change, while the input (mains) voltage is held constant.

Both control characteristics can be given either as a percentage of the output voltage, or directly in

Table 4.4 Polarity and range limit assignment on the Kepco SN-488-B program card

ASCII character	*Converted binary value*	*Output A* *Polarity*	*Output A* *Amplitude*	*Output B* *Polarity*	*Output B* *Amplitude*
0	0000	+	H	+	H
1	0001	−	H	+	H
2	0010	+	L	+	H
3	0011	−	L	+	H
4	0100	+	H	−	H
5	0101	−	H	−	H
6	0110	+	L	−	H
7	0111	−	L	−	H
8	1000	+	H	+	L
9	1001	−	H	+	L
A	1010	+	L	+	L
B	1011	−	L	+	L
C	1100	+	H	−	L
D	1101	−	H	−	L
E	1110	+	L	−	L
F	1111	−	L	−	L

engineering units (mV). In the specification the degree of input (mains) voltage or load change (that causes the output voltage change) is also given.

Ripple and noise voltage: This is the value of alternating voltage superimposed on the output direct voltage. It can be given either as an effective or as a peak value, relative to a certain frequency band.

Isolation voltage: This is the maximum voltage that can be sustained safely between the output points of the power supply and ground. A knowledge of this is especially important for power supplies connected in series.

Programmable current limit: 'Current limiting' is an overload protection method. In most programmable power supplies the value of the current limit and the value of the output voltage are changeable (Figure 4.13).

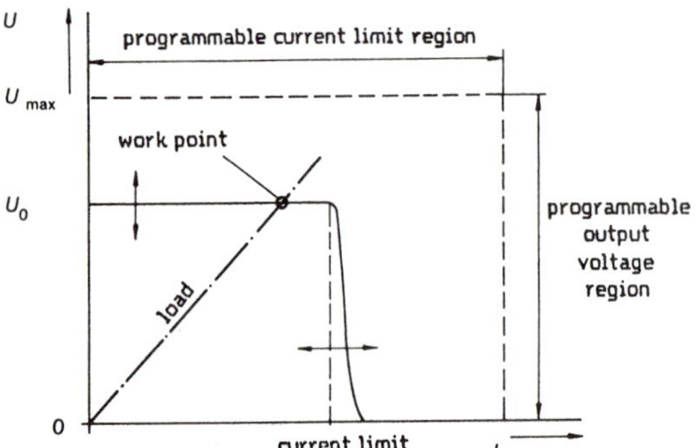

Figure 4.13 Characteristics of a programmable power supply unit

Programmable overvoltage protection (crowbar): Overvoltage protection is an automatic voltage-limiting method. It operates on the basis that a low value, parallel resistance is connected to the output, should the output voltage exceed a predetermined limit. The value of the limit can be programmed externally in the programmable power supplies.

Calibration: Under ideal conditions the output voltage of voltage-programmable power supplies is directly proportional to their control voltage (Figure 4.14(a)), i.e.:

$$U_{out} = \frac{U_{out}}{U_{prog}} U_{prog} = mU_{prog}$$

where U_{out} is the output voltage of the supply source unit, U_{prog} is the voltage controlling the programming of it, and m is a constant.

In reality, however, zero value output voltage does not correspond to zero control voltage (Figure 4.14(b)). The calibration is basically the elimination of this offset, which is usually of a few millivolts magnitude only.

Programming speed: The controlling speed of programmable power supplies is determined mainly by the output capacity and the load impedance of the power supply unit. There are two different situations, according to whether a decrease or an increase of the supply unit's output voltage is required.

The equivalent circuit and the output voltage–time function relevant to voltage increase are shown in Figure 4.15(a). The control circuit of the power supply unit detects at time t_1 that the output voltage is smaller than the required value. Then the charging of the RC component – consisting of the

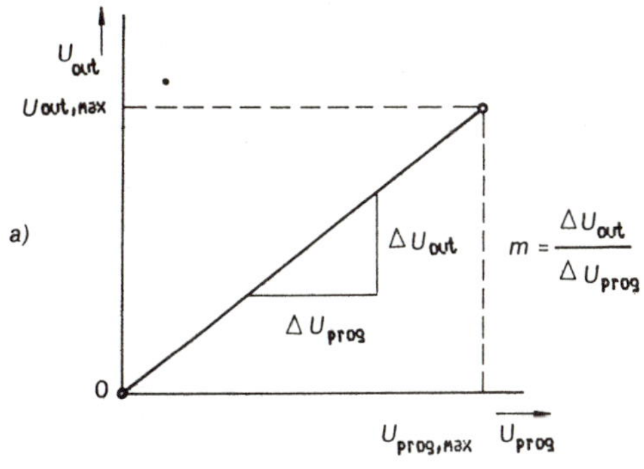

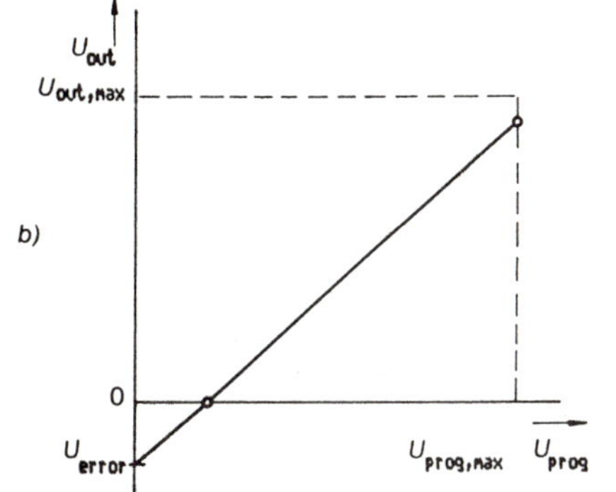

Figure 4.14 Relationship of the control and output voltages of a programmable power supply unit. (a) In theory, (b) voltage decrease

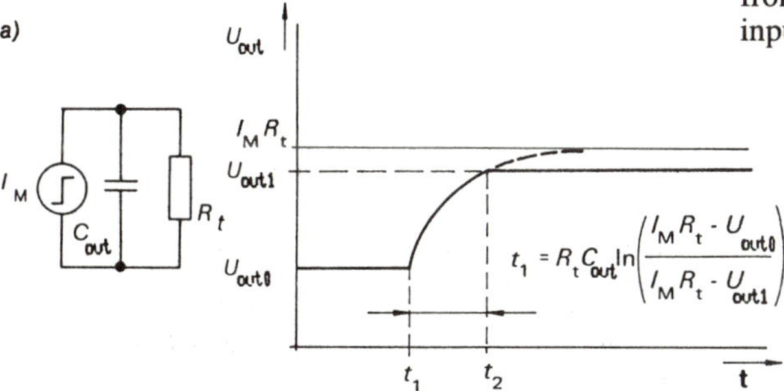

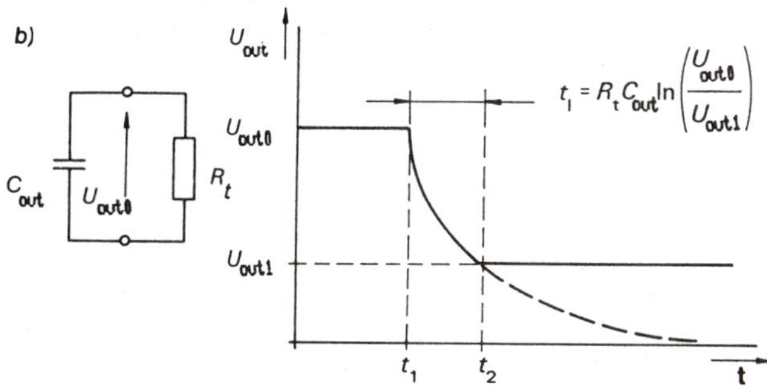

Figure 4.15 Equivalent circuit of the programmable power supply unit output in dynamic operational mode. (a) Voltage increase, (b) voltage decrease

output capacitor (C_{out}) and the load resistance, R_t – commences, using the maximum current, I_M allowed by the serial controller of the power supply unit. The output voltage inceases exponentially at a rate determined by the time constant, R_tC_{out}. According to the method shown in the Figure, the response time, t_f, is determined by the value of the voltage change and the permitted maximum current and is limited by the time constant. The response time has a minimal value if there is no load present, when the output voltage increases linearly.

The equivalent circuit and the output voltage–time function relevant to voltage decrease are shown in Figure 4.15(b). In this case the output voltage also decreases to the value required by the program control at a rate determined by the time constant R_tC_{out}.

It is apparent from the above that one way of 'accelerating' the response of the power supply unit is to decrease the output capacity. However, this is detrimental to noise reduction and is not acceptable in most cases; in any case, with increasing voltages the operation can be accelerated by increasing the magnitude of the current limit.

4.3 Measurement-point switches and relays

In many measuring systems the signal sources and the instruments cannot be connected directly to the measured unit. The reason for this can originate from the need to transmit the same signal to several input points, or to use the same instrument at several measuring points. The converse of this situation applies when various instruments must be connected to the same point of the measured unit in consecutive measuring cycles. If the control of the instruments and the processing of data are carried out automatically while continuing to use manual measuring-point switching, the most important advantages of automation (fast operation, elimination of subjective, human errors, etc) are lost. Hence, electronically-controlled measuring-point switches or relays (otherwise known as 'matrix switches') were used in even the earliest automatic measuring systems.

Following the acceptance of the IEC interface system, the development of this field was accelerated and new products are almost always IEC-compatible.

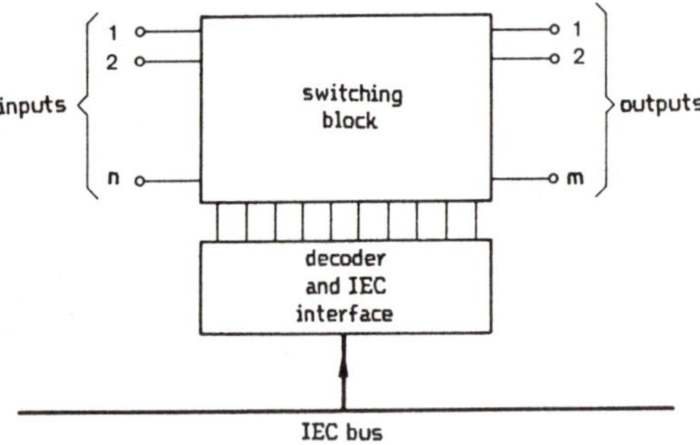

Figure 4.16 Structure of the IEC-compatible measuring-point switches

Figure 4.16 shows the general block diagram of IEC-bus-controlled measuring-point relays. The most important component of the device is the switching block, containing the switching elements connecting the input points to the output points under the direction of the controller. The structure and the organization of the switching block, as well as the type of relays used will determine the tasks it will be able to carry out. The tasks of the device's other functional elements are the generation of the IEC interface listener and receiver functions, program decoding and switching element controls.

4.3.1 Classification of measuring-point relays and switches

Measuring-point relays and switches can be classified in many ways. In the following section various categories and their most important characteristics will be described.

4.3.1.1 Switching elements

There are two main categories of switching element used in measurement-point switching assemblies: these are 'mechanical relays' and 'semiconductor switches'.

Mechanical relays: The most important mechanical relays are 'reed relays'. The contacts of reed relays are enclosed in a hermetically-sealed glass container and are operated by permanent or electromagnets located outside the glass. Owing to the protected contacts, the operation of reed relays is very reliable, their life time being typically about 10^6 to 10^7 operations.

One of the major advantage of reed relays over semiconductor switches is that their transmission (contact) resistance is small (5 to 100 mΩ), although in the open state their isolation resistance is practically infinite, due to the physical separation of the contacts. The isolation resistance towards other channels and the control unit can also be regarded as infinite. The range of voltage that can be switched is very large, i.e. 50 to 300 V.

The disadvantage of reed relays is comparatively slow operational time (0.5 to 5 mseconds), which is determined by the rise and decay of the magnetic field and particularly by bounce of the contacts.

There are two special versions of reed relay that are commonly used in addition to the above. One of these is the so-called 'mercury-wetted relay', which has no contact bounce and has a much longer life (10^9 operations).

Another kind of special reed relay is the 'low-thermal reed', which is used in low-level signal circuits. While the thermoelectric potential between the iron reed contact and the copper connection is around 40 μV in normal reed relays, this error potential is only 1 to 5 μV for the low-thermal type.

Semiconductor switches: Of the various semiconductor switches the field effect controlled transistor (FET) is the most generally used. All three basic types of FET (JFET, MOSFET and CMOS) are used as switching units; however, the advantages of the CMOS switches (especialy the very small current drain and the very large voltage switching range) are particularly attractive in many applications.

The advantages of the semiconductor switches over mechanical relays are their practically unlimited life and very short switching time (50 to 1000 nseconds). However, they suffer from the disadvantage that their transmission (contact) resistance is much higher (20 to 100 Ω) and they have some transmission even in the open state, where their isolation resistance is more limited (10^6 to 10^9 Ω). Their major disadvantage is their very limited switching voltage range, which in practice is determined by the supply voltage; furthermore, they can be damaged very easily by the application of excessive voltages and require protection against these.

4.3.1.2 *Multiplexer structure*

Relays can also be classified according to the structure of the switching block. From this aspect there are three categories.

Multiplexer units or *scanners* have two or more data inputs. The input to be connected to the output is selected by a signal combination switched onto the program inputs (Figure 4.17(a)).

The task of *demultiplexer units* is the opposite, the output point to be connected to the input signal is selected by the signal combination switched onto the program inputs (Figure 4.17(b)).

Matrix switches are the third category of measuring-point switches. These units can connect any input point to any output point, according to the controller's instructions (Figure 4.17(c)).

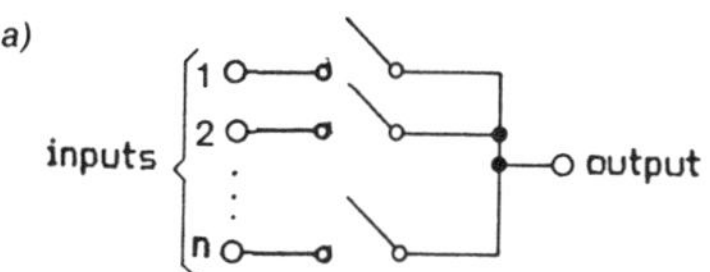

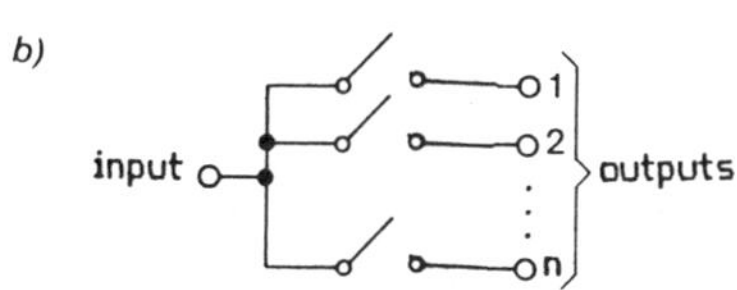

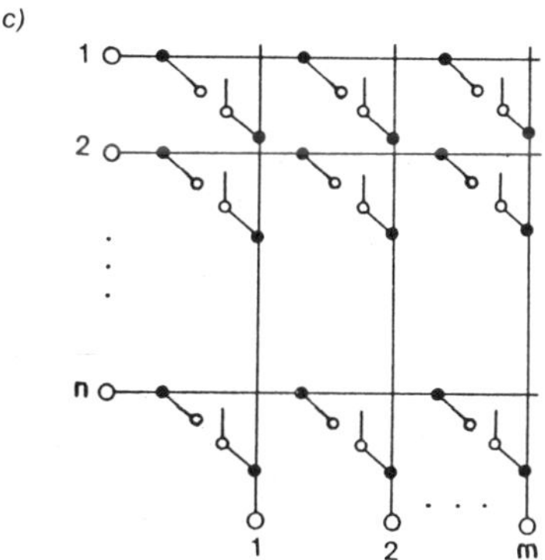

Figure 4.17 Classification of measuring-point switches. (a) Multiplexers, (b) demultiplexers, (c) matrix switches

4.3.1.3 *Switching sequence*

From the switching sequence aspect there are arbitrary or fixed sequences. Arbitrary-sequence measuring-point switches always contain a decoder unit that generates the signals operating the relay units from the program data.

Fixed-sequence measuring-point switches require only a start signal and a relay timing signal for their operation, as the switching sequence of the unit cannot be changed.

The *cyclic scanner* is a special form of fixed sequence multiplexer and is almost always used for data collection.

4.3.1.4 *Frequency band*

Measuring-point switches can also be classified according to the frequency band being switched.

In the construction of switching blocks of measuring-point relays used for low-frequency signals the most important aspect is to provide the best possible protection against interference. The transmission of millivolt signals, especially over large distances, cannot be achieved without shielding. In many cases simple shielding is insufficient and the use of guarding will be necessary to provide protection from in-phase interference signals. This means that three contacts will be necessary for switching one channel. Hence, the basic unit of low-level, low-frequency signal measuring switches normally contains three reed relays (Figure 4.18). Two of these are low-thermal reeds, whilst the third, switching the shielding, is a normal reed. The contacts switching the shielding are opened last at switch-off and closed first at switch-on.

High-frequency measurement relays must be used for the undistorted transmission of fast rise time impulse signals, as the > 100 MHz frequency spectrum components of such signals is significant. However, the frequency limit above which these relays must be used cannot be defined precisely, as the choice is influenced by the length of the connections used in the system, as well as the frequency of the measured signals.

For the connection of high-frequency signals transmission lines are used. One of the most important features of such lines is the characteristic impedance, determined by the geometric dimensions of the line. If the transmission lines are not terminated with the characteristic impedance, or if the characteristic impedance changes on the line for some reason (e.g. connection or switching), detrimental reflections will occur leading to distortion of the transmitted signals. Hence, in using high-frequency measuring relays, it is very important to pay special attention to line impedances and terminations. Stray capacities play an important part in determining the characteristic impedance of the line, since their effect is no longer negligible at high frequencies.

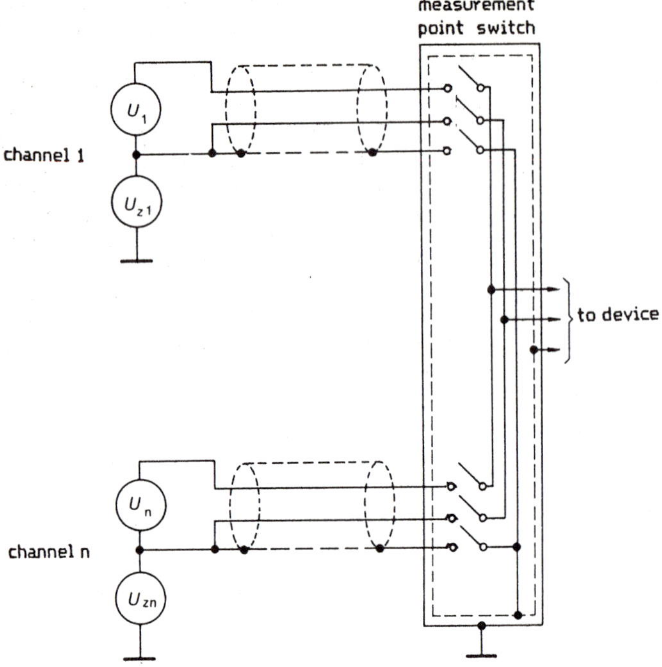

Figure 4.18 Use of the guarding technique in low-frequency measurement-point switches

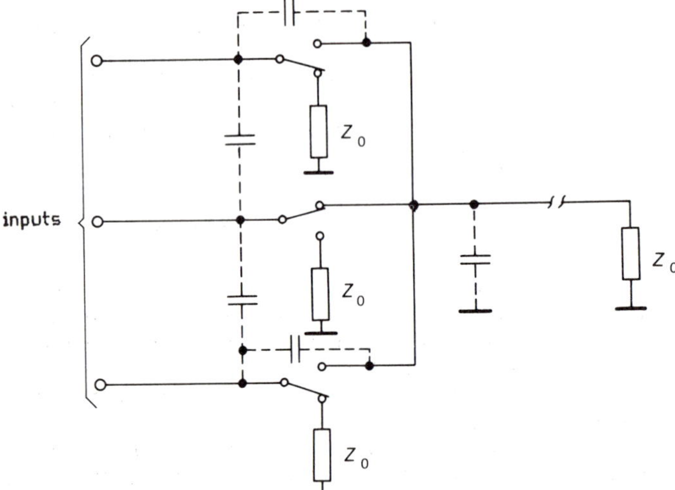

Figure 4.19 Equivalent circuit of a multiplexer including stray capacities

Figure 4.19 shows the theoretical circuit diagram of a high-frequency multiplexer. The inputs are always terminated with a characteristic impedance (Z_0) independent of the on/off state. The diagram shows the stray capacities that reduce the isolation resistance in the off state and cause insertion losses in the on state. They contribute to interference, or crosstalk, in both states. The insertion loss and the interference caused by the stray capacities are directly proportional to the frequency.

4.3.2 Characteristics

The function of measuring-point switches is seemingly very simple, i.e. according to a program they connect various points of the measured unit with the signal sources or with the instruments. However, in practice the measuring possibilities of an automatic measuring system are often limited by problems arising from these switches.

The ideal measuring point switch has a zero

transmission resistance, an infinite isolation resistance and zero switching time. As we have seen, real devices do not possess these idealized characteristics and consequently influence the operation of the system connected to them to a greater or lesser degree. Let us examine the characteristics that influence the choice of a suitable measuring-point switch for a given measuring task.

4.3.2.1 Characteristics of low-frequency measurement point switches

Configuration: The most important characteristic of a measuring-point switch is the internal configuration of the switching block, the number of inputs and outputs and the structure of the basic switching unit. The structure of automatic measuring systems is varied and the modern measurement-point relay is invariably modular. The contact system of the relays is usually given using the internationally-accepted abbreviations. These symbols are listed in Table 4.5.

Table 4.5 Relay contact symbols

Symbol		
	A S.P.S.T.– N.O.	Form A terminating contact
	B S.P.S.T.– N.O.	Form B disconnecting contact
	C S.P.D.T.	Form C Morse contact
	D D.P.D.T.	Dual Morse contact

Maximum contact ratings: The switching contacts are specified by the maximum voltage and current capacities that can be handled.

Relay life: The life time is given by the number of operations executed under fully-loaded conditions, before failure.

Contact resistance: The value of this is calculated from the voltage drop generated by the highest permitted current on the closed contact.

Insulation or isolation resistance: The value measured on the open switch using a direct test voltage of 100 to 500 V.

Switching time: In the case of programmed-control measuring-point relays the 'switching time' means the time required for the complete switching operation, starting from the reception of program data and ending with the contacts arriving to their stable position. This is one of the most important characteristics of relays used in automatic measuring systems. Its value directly influences the system's operational speed, as is apparent from Figure 4.20, which shows the timing relationships of a calculator controlled data collecting system containing a scanner and a digital voltmeter.

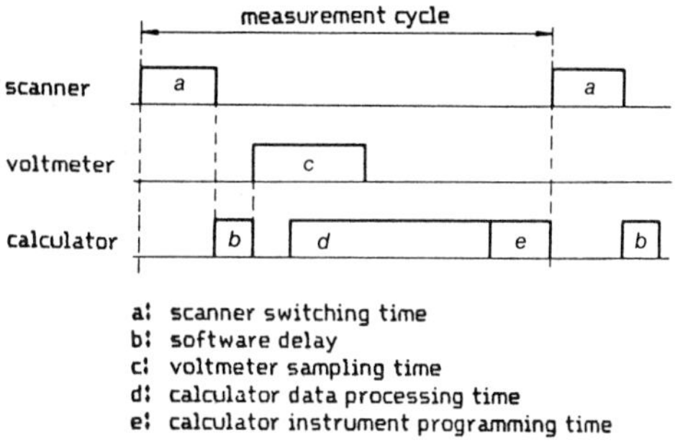

Figure 4.20 Timing relationships of a data-collecting system

Crosstalk: This is a measure of the interference from adjacent channels. Its value depends mainly on the frequency and the proper terminations of adjacent channels.

4.3.2.2 Characteristics of high-frequency measurement switches

Characteristic impedance: This refers to the internal circuit and the switching device of the measurement-point relay, providing a strict standard for the cables connected to it. In practice the 50Ω-characteristic impedance is the most widely used.

Voltage standing wave ratio (VSWR): VSWR is a measure of the reflections due to the insertion of the measuring-point relay. Its value is greatly dependent upon frequency (Figure 4.21).

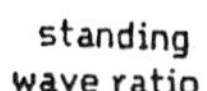

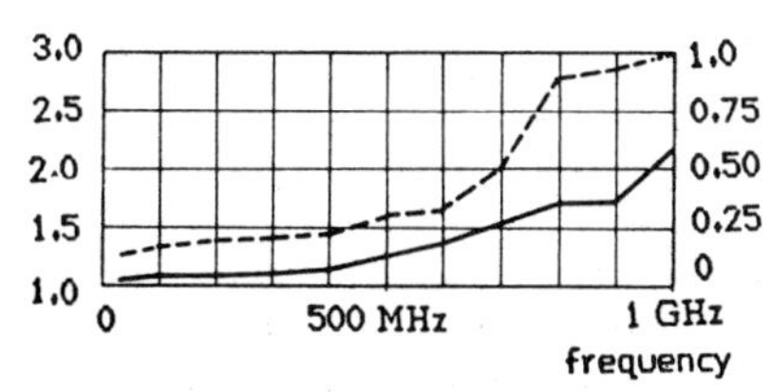

Figure 4.21 Standing wave ratio and insertion loss of a high-frequency measurement-point switch as a function of frequency (E-H Research, Type 42)

Insertion loss: This is the value of the voltage drop, expressed in dB units, due to the insertion of the measurement-point relay. It is also dependent upon frequency (Figure 4.21).

Rise time: In the case of high frequencies the stray capacity of the measurement-point relay, the loss resistances and the cable inductance form a passive quadrupole. The rise time of this increases the rise time of the transmitted impulse.

4.4 Modular universal devices

Nowadays the majority of the IEC-system-compatible instruments are of traditional structure, i.e. they can be controlled from either the IEC bus (excluding the local handling devices) or used as self-contained instruments utilizing the front panel handling devices and displays. However, should the user require the instrument for automatic system applications only, then he is reluctant to pay extra costs to provide built-in local controls that are never used.

Recognizing this obvious fact, more and more instrument manufacturers are developing automatic instrument sets that contain generally used instruments (multi-meter, frequency meter, transient recorder, etc) and are suitable for varied user applications due to their modular structure.

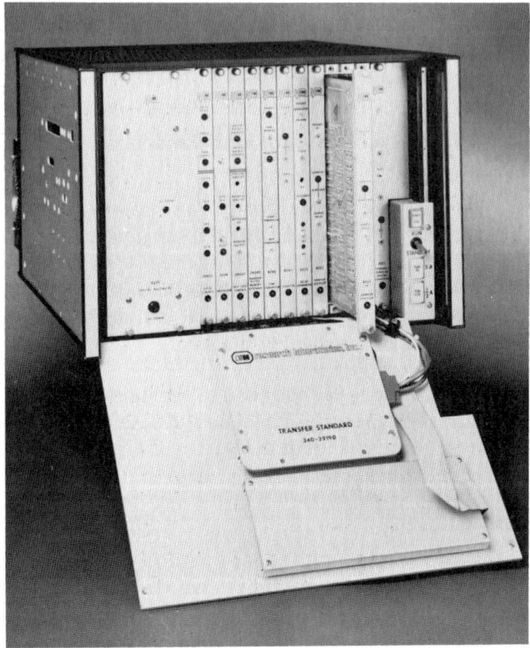

Figure 4.22 Modular E-H Research Type 8200 measuring system

An important advantage of these systems is that their structures can be changed. The importance of this can be appreciated, if we consider that the useful life of an instrument is determined by the very rapid rate of technical obsolescence that is now common. Such systems can be readily extended by the user at low additional cost.

Figure 4.22 shows an example of such a universal measuring system, the Type 8200 made by E-H Research. This device is a plug-in instrument set whose structure can be changed and the individual elements of which can be operated using a controlling microprocessor or IEC interface.

The structure of the 8200 system is shown in Figure 4.23. The individual units belong to three categories. The units handling the control function are the microcomputer unit together with the memory expansion and the IEC interface unit directly connected to it, the matrix switch and the validation generator necessary for automatic calibration.

The basic measuring units are the frequency meter, the clock, the digital delay unit and the digital multimeter, as shown in the Figure. Three slots are provided in the frame of the device to accommodate auxiliary units.

Controller part: All measured and trigger signals are transmitted to the input of the appropriate instrument via the matrix switch, which is controlled by the microcomputer unit according to the program instructions. The central unit of the microcomputer is a Type MC-6800 microprocessor, located in one of the slots. The operational system controlling the operation of the measuring system is located in the 8-kbyte ROM, and the user's programs and measurement data are stored in a 2-kbyte RAM. The latter can be extended to a maximum of 32 kbytes, using an extension card. Apart from the instrument control, the microcomputer can be used for automatic calibration of the entire system, using the built-in validation generator.

The operational language of the system is a very simple mnemonic program language and the programming of the instrument requires a three-letter code. A controlling character string can be the following, for example:

DMM,FUN=VDC,RAN=6,MOD=DIF,

where DMM is the instrument allocation (digital multimeter), FUN=VDC is the operational mode allocation (direct voltage measurement), RAN=6 selects the range (1000 V), MOD=DIF is the trigger mode and filter option (direct, filtered).

The program language also contains 19 two-letter control commands which are used for program construction and for saving in memory.

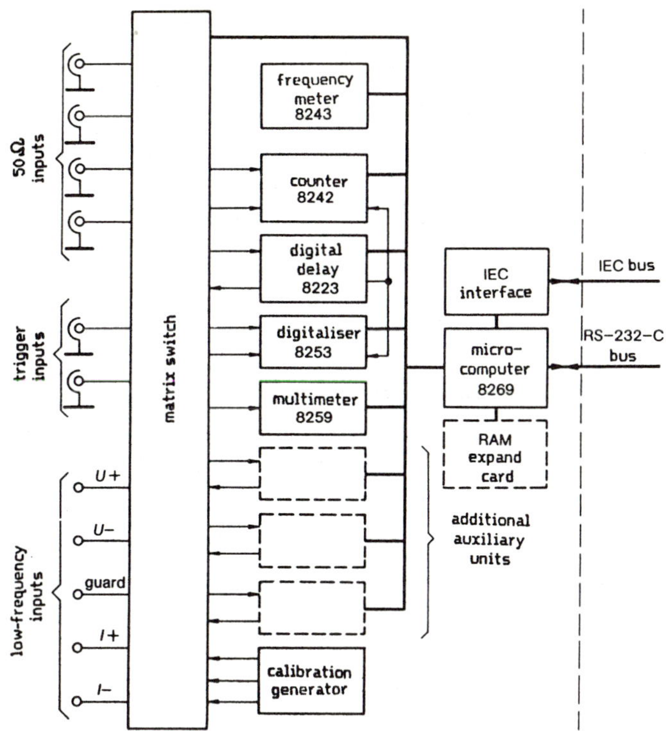

Figure 4.23 Structure of the E-H Research Type 8200 measuring system

From the IEC system aspect, the 8200 system is a device possessing talker and listener capabilities. Apart from the IEC interface (used in the on-line operational mode) the 8200 system also has another interface, allowing the connection of any RS-232-C compatible computer peripheral. Such a device can be used for program construction and calibration in the off-line operational mode.

Measuring units: The slot units of the 8200 measuring system are controlled by the micro-processor unit via the internal bus. The individual slots possess basically-independent functions, although the designer has tried to make good use of the advantages arising from the integrated structure, where possible; for instance the inputs of the clock and the frequency meter are shared.

The basic measuring units belong to two categories: (1) The Type 8253 digitizer (which is a 100-MHz transient recorder) and the Type 8259 multimeter are instruments that measure signal amplitude characteristics. (2) The Type 8243 counter and the Type 8242 counter measure the characteristics related to time functions of the signals.

Connecting the two instrument groups is the Type 8223 digital delay unit which delays the start of the further measuring units in 1 nsecond steps up to a maximum of 10 mseconds.

Similarly, modular measuring systems are manufactured by other companies. The most well-known of them is Hewlett-Packard's Multiprogrammer, a recent version of which (Model 6942A) offers simplifed programming and minimal loading of the controller. A British firm BIODATA of Manchester produces a highly-flexible system which can be configured to suit the most diverse user demands. Their MICROLINK family has a constantly-expanding range of modules.

4.5 IEC interface units

Due to the universal acceptance of the IEC system, many of the instruments designed in the last few years contain the IEC interface as a built-in optional unit. However, there are still many instruments on the market which are without the IEC interface. Hence, many companies manufacture interfaces that allow the connection of the instruments' traditional BCD interface output and the IEC bus. These interface units can be classified according to their flexibility, i.e. whether they can be used for interfacing a single instrument type, or for various instruments, i.e. they are universal.

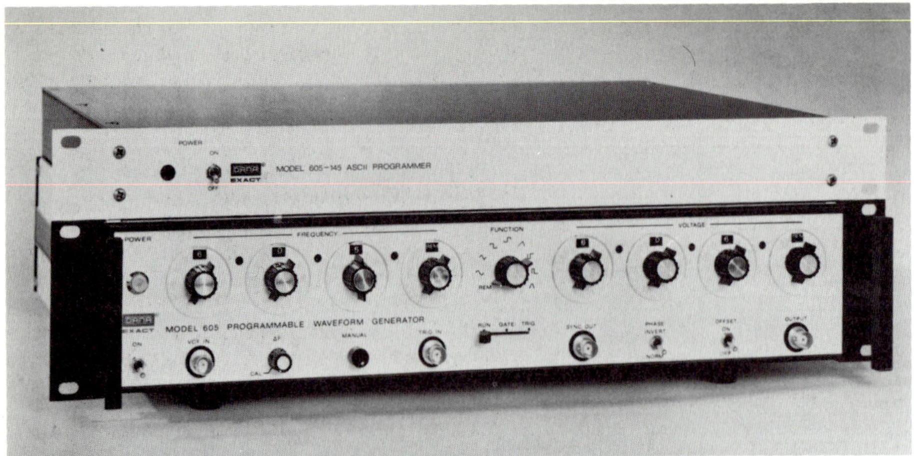

Figure 4.24 Model 605-145 programmable waveform generator made by EXACT, with added IEC interface

The Type 605-145 interface, made by EXACT, is a specific unit, interfacing the EXACT Type 605 waveform generator to the IEC bus (Figure 4.24). The waveform generator is controlled via a 40-pin connector located on its back. The numbers of inputs providing control of the various functions and the applicable codes are listed in Table 4.6.

The interface generates the device controlling parallel codes from the ASCII coded serial character string arriving from the IEC bus. In the program table of the interface the IEC bus ASCII characters are allocated specifically to the device functions.

A similarly unambiguous connection cannot be given for general purpose IEC interfaces. Figure 4.25 shows the ICS Electronics Type 4881 and Type 4882 interfaces and the connection of the instruments interfaced to them. Both interfaces are entirely general purpose, i.e. they can be used for interfacing any make and type of instrument possessing BCD input or output, provided that certain basic conditions are met.

The listener device of the Type 4882 interface can receive a 12-character string message from the IEC bus, after having received its address. The messages must always be terminated with the CR LF characters. The 10 places preceeding the delimiter can contain the following ASCII characters:

123456789.;+−E

After the reception of the delimiter, the Type 4882 interface puts the device-dependent message into the output register and then transfers it in parallel format to the instrument simultaneously with the positive going edge of the TP (transfer control) impulse. If the instrument is not yet ready to receive the data, this is signalled to the interface

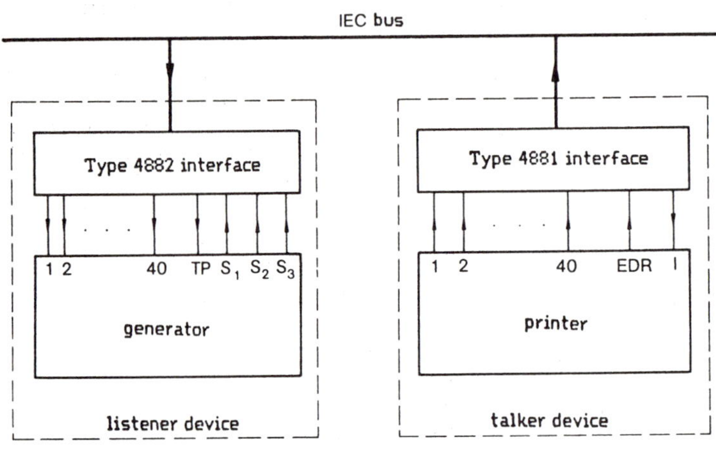

Figure 4.25 Connection of the ICS Electronics Type 4881 and Type 4882 interfaces

Table 4.6 Configuration of the Type 605 EXACT Function Generator controls

Function	*Number of inputs*	*Code*
Frequency limit	4	BCD
Frequency set	12	BCD
Voltage limit	2	binary
Voltage set	12	BCD
Signal shape	4	binary
Polarity	1	binary
Fix DC offset	1	binary
Trigger	2	binary
Common ground	1	–

via the S_1, S_2 and S_3 state flag lines. This influences the handshake process of the interface, as new program data cannot be transmitted until the exchange of the first one has been completed.

The Type 4881 interface operates on similar principles and can be connected to instruments requiring talker functions. From the 4-bit BCD outputs of the device ten characters can be transferred to the input registers of the Type 4881. The data transfer is triggered by the positive going edge of the EDR (exit data ready) validation signal. The interface signals the instrument with the I (inhibit) signal that the received device-dependent message was transmitted to the IEC bus and it is ready for the reception of the next message.

The values of the IEC bus transferrable characters are limited by the 4-bit BCD data input in the case of the Type 4881, as shown in Table 4.7.

Table 4.7 Code table of the ICS Electronics Type 4881 interface

Decimal value of the 4-bit BCD input character	*ASCII code of output character*
0...9	0...9
10	+
11	–
12	.
13	,
14	E
15	△

Similar interfaces are produced by Farnell Instruments. Their Omnibus model OB1 performs both the talker and listener functions of the IEC bus, while Model OB2 has identical specification with the addition of an integral 3½-digit panel meter.

In comparing the capacities of specific or universal IEC interfaces, the limited character sets of the latter is evident. The universal interfaces are suitable for transferring numeric data only, and in most cases this is sufficient only for simple IEC systems.

4.6 IEC interface circuits

Figure 4.26 shows the structure of an IEC measuring system. The IEC interface connected units of one of the devices, an intelligent instrument, are drawn in detail. The logic part of the IEC interface shown in the Figure can be incorporated in a single chip and suitable LSI interface circuits are marketed by an increasing number of semiconductor manufacturers. In the Chapter discussing microcomputers, the Type 8291 circuit interfacing the 8-bit Intel microprocessors to the IEC bus has been mentioned already. It is used in talker and listener devices. Similar purpose instrument interfacing circuits are produced by Motorola (Type MC68488), the NEC (Type 7210) and by Texas Instruments (Type TMS9914) for their microprocessors.

There are special buffers for connecting the IEC chips to the bus lines. A typical circuit is the National Semiconductor Type DS3666, which is a high-speed Schottky 8-channel bi-directional transceiver.

In addition Philips produce a Type HEF4738V universal LSI interface circuit that is suitable for the interfacing of any microprocessor or wired control instrument and the IEC bus. These interface units greatly simplify the microprocessor software interfacing section, thus making the task of instrument control much easier.

In the following sections the structure, the operation and the use of the Motorola and the Philips interface circuits will be discussed in detail.

4.6.1 The Motorola Type MC68488 interface circuit

The Type MC68488 general purpose interface adaptor is an NMOS technology LSI circuit, built into a single 40-pin ceramic chip. The unit and some auxiliary circuits generate the following interface functions when connected to the Motorola Type MC6800 or Type MC6802 microprocessors:

receiver and source handshake
talker or extended talker
listener or extended listener
service request
remote/local control
parallel poll
device clear
device trigger.

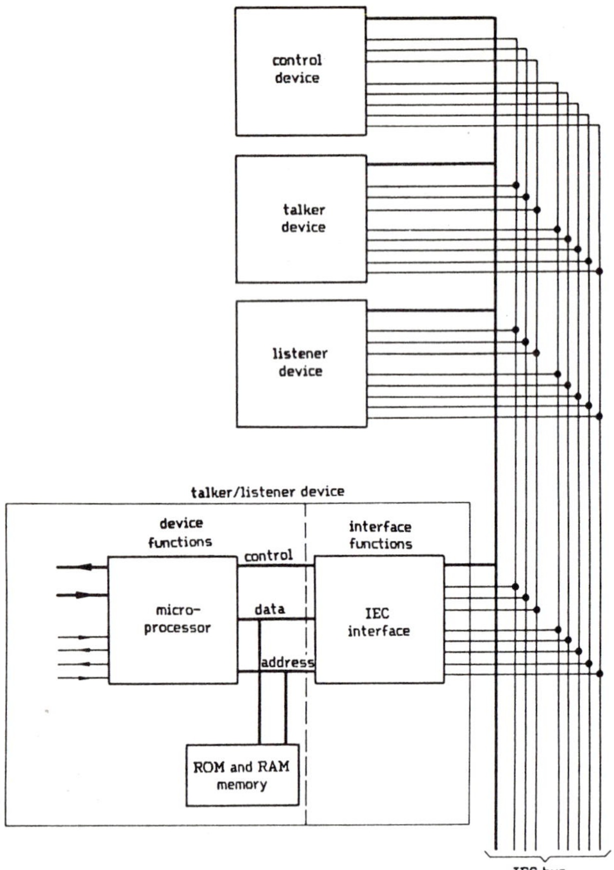

Figure 4.26 Interfacing a programmed-control instrument to the IEC bus

Apart from the above interface functions the circuit can also generate some device functions, e.g. address recognition, DMA control, 'talk only' and 'listen only'. The generation of the various functions is either automatic in the interface unit, or achieved by a minimal amount of microprocessor usage.

4.6.1.1 *Input/output signals*

Figure 4.27. shows the interfacing of the MC68488 unit to the IEC bus or the microprocessor.

The connection of the IEC bus and the interface unit's signal lines is achieved by four Type MC3448A non-invertible bus receiver/transmitter circuits. Each of these circuits is suitable for the control of the signal lines and they can operate either as open-collector or as three-state drivers, according to the external control signal. This selection and the data traffic of the entire IEC bus are controlled by the TR1 and TR2 signals of the interface. Although the receiver/transmitter units are non-inverting, the negative values of the IEC bus signals are used for the assignment of the MC68488 outputs. This is because the IEC interface uses a negative logic convention, while the Series MC68XX Motorola circuits use positive logic.

The two-way data traffic between the MC68488 and the microprocessor takes place on the DB0...DB7 data buses, via three-state drivers. The control signals handle the following functions:

- RS0, RS1, RS2 register selecting signals; their task is, combined with the R/W (read/write) signal, to select the register from the 15 available on the MC68488 for writing and reading.
- CS chip selection signal, which is used by the microprocessor for the selection of the MC68488 unit from the various possible peripheral units.
- CLK clock signal, which is the 02 clock signal of the microprocessor and has a maximum value of 1 MHz.

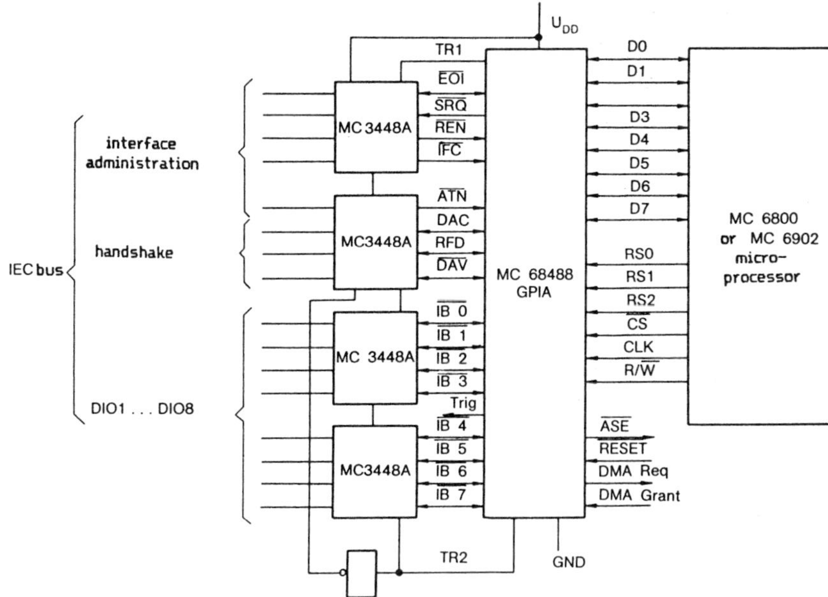

Figure 4.27 Interfacing a microprocessor and the IEC bus via the Motorola MC68488 circuit

- R/W (read/write) signal, which is used by the microprocessor for the control of data traffic direction.
- ASE address switch enable, which is the enable signal of the three-state drivers separating the device address set switches.
- Reset clearing signal, which is used for restoring the MC68488 unit to ground state after the power supply has been switched on.
- DMA Req, DMA Grant, DMA control signals; their task is to ensure the mutual operation of the MC68488 and the DMA units in the DMA operational mode.

4.6.1.2 Structure and operation

The MC68488 interface unit contains 15 8-bit registers which are used for the transfer of data to the IEC bus or the transmission of data or messages arriving from the bus to the microprocessor. The contents of the registers are compiled in Table 4.8. Only seven of the registers can be written to (RXW) and eight of them can be read from (RXR). The write registers are filled from the microprocessor, while the read registers contain mainly IEC interface commands and addresses. The various state signals (flags) generated by the interface are also contained in the read register.

In the talker addressed state of the device the data are transferred to the 'Data out' (R7W) register, then to the IEC bus, using a handshake process. If all devices belonging to the bus accept the data, then the BO (byte output) bit of the 'Interrupt State' register signals the microprocessor that the data is available.

In the extended talker/listener functions the microprocessor can read the secondary device address from the 'Command Transfer' register (R6R). If the microprocessor accepts the secondary address as its own, then it transmits the 'My Secondary Address' (MSA) message via the 'Auxiliary Command' (R3W) register to the IEC bus.

The recognition of the primary address takes place in the interface. The allocation of the address can be either in the program or by using the address switches and three-state drivers, as shown in Figure 4.28. At the allocation of the address the $\overline{ASE}$ control signal gives permission for the input of the information set on the address switches to the microprocessor. Then the microprocessor writes the address in the 'Address' (R4W) register.

The interface unit responds to the IEC system controller originated polls via the 'Serial poll' (R5W) and the 'Parallel poll' (R6W) registers.

The MC68488 is capable of generating the 'talk only' and 'listen only' messages, as well as the standard IEC interface functions. These two states are assigned by the microprocessor with the 'to' and 'lo' bits contained in the 'Operational Mode' (R2W) register.

Table 4.8 The MC68488 interface registers

Register name	*Symbol*	*Bits*							
		7	6	5	4	3	2	1	0
Interrupt	R0W	IRQ	BO	GET		APT	CMD	END	BI
Interrupt state	R0R	INT	BO	GET		APT	CMD	END	BI
Command state	R1R	UACG	REM	LOK		RLC	SPAS	DCAS	UUCG
Not used	R1W								
Address state	R2R	ma	to	lo	ATN	TACS	LACS	LPAS	TPAS
Operational mode	R2W	dsel	to	lo		hlde	hlda		apte
Auxiliary command	R3R R3W	Reset	DAC rfdr	$\overline{\text{DAV}}$ feoi	RFD dacr	msa	rtl	ulpa dacd	fget
Address switch	R4R	UD3	UD2	UD1	AD5	AD4	AD3	AD2	AD1
Address	R4W	lsbe	dal	dat	AD5	AD4	AD3	AD2	AD1
Serial poll	R5R R5W	S8	SRQS rsv	S6	S5	S4	S3	S2	S1
Command transfer	R6R	B7	B6	B5	B4	B3	B2	B1	B0
Parallel poll	R6W	PPR8	PPR7	PPR6	PPR5	PPR4	PPR3	PPR2	PPR1
Data in	R7R	D17	D16	D15	D14	D13	D12	D11	D10
Data out	R7W	DO7	DO6	DO5	DO4	DO3	DO2	DO1	DO0

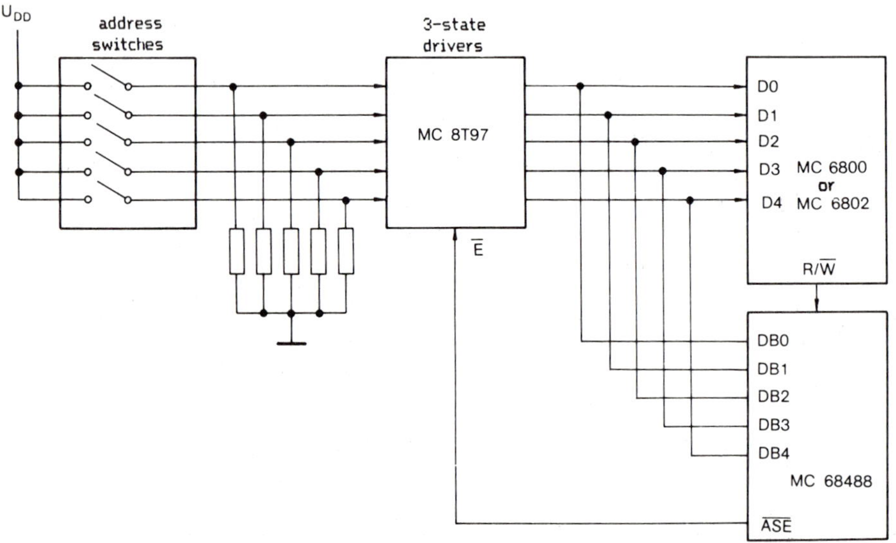

Figure 4.28 Device address allocation in the MC68488 interface unit

4.6.2 The Philips Type HEF4738V interface unit

The Type HEF4738V universal interface unit is a LOCMOS (Local Oxidation Complementary MOS) technology 40-pin DIP LSI integrated circuit. LOCMOS is an advanced form of CMOS and its main advantages are smaller chip size for the same number of functions, operational speed approaching that of TTL technology and wide supply voltage range (3 to 15 V).

When connected to a few simple auxiliary circuits the HEF4738V can generate the following interface function variants:

receiver and source handshake (AH1, SH1)
talker (T1 or T5)
listener (L1 or L3)
service request (SR1)
remote/local control (RL1)
parallel poll (PP1)
device clear (DC1)
device trigger (DT1).

Using the HEF4738V the interface related tasks of the device (e.g. instrument) are greatly simplified. For the generation of the above functions only two listener and two talker handshake lines, three so-called 'device command lines' and a few data-transfer lines are necessary between the HEF4738V and the device. The interface unit and the IEC bus are connected via inverting receivers/transmitters.

4.6.2.1 Structure and operation

The structure of the HEF4738V is shown in Figure 4.29. The inverted data lines of the IEC bus are connected to the message decoder and address comparator unit. As the interface messages arriving from the bus are transmitted on the DIO1...DIO7 lines, the DI08 line is not connected to the interface unit. The messages have negative values on the IEC bus lines of the interface unit that is using positive logic values, with an additional 1 (input) or 0 (output) signal. The control and data transfer signals of the IEC bus are transferred to the appropriate functional units of the interface. The interface generates the messages sent to the device and the signals of the two handshake processes from these control and data transfer signals and from the decoded interface messages.

4.6.2.2 Local messages

The decoding of the local messages arriving from the device is the task of the identifying register of the interface unit. This step register is filled through a single input (lsr), thus saving ten input points.

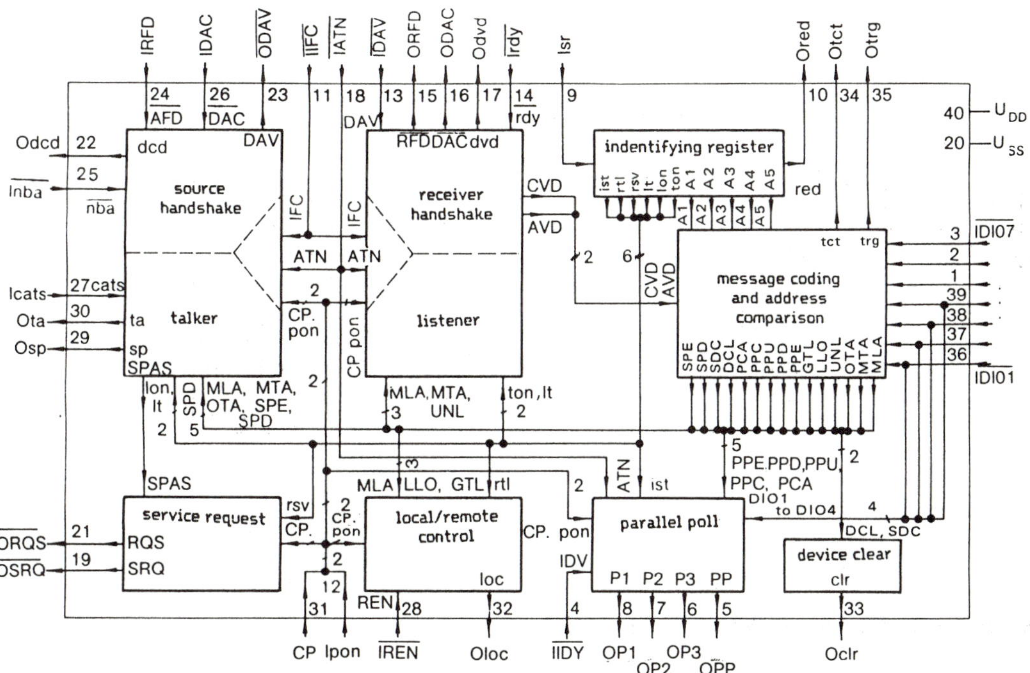

Figure 4.29 Circuit diagram of the Type HEF4738V Philips interface circuit

However, an external auxiliary circuit is necessary with this solution. The serial input of the local messages can be achieved by using two 8-bit step registers, e.g. HEF4014B, as shown in Figure 4.30. The optimum frequency of the clock signal that is stepping the registers is 2 MHz. The contents of the identifying register are continuously updated, according to the actual values of the local messages. In this continuous process the OR signal has an important role in permitting the parallel filling of the external step registers prior to the serial filling of the identifying register.

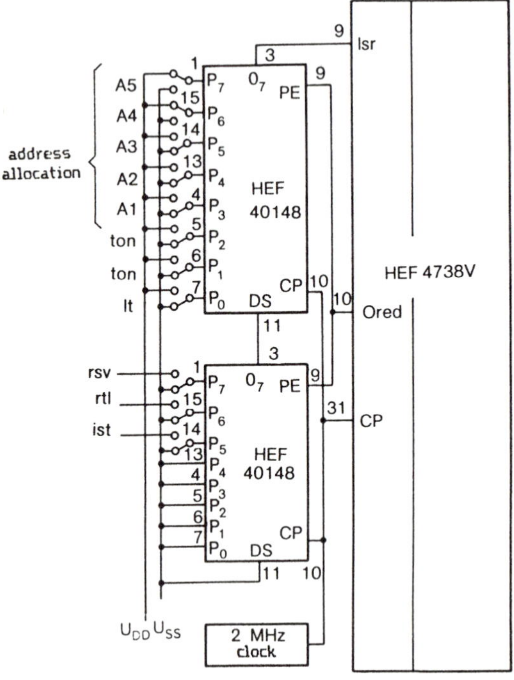

Figure 4.30 Input of local IEC interface messages into the HEF4738V interface circuit

The information stored in the identifying register corresponds to the following local messages:

A1...A5	device listener and talker address
ton	the 'talk only' message
lon	the 'listen only' message
lt	the section of the talker and listener variant
rsv	the 'request service' message
rtl	the 'return to local control' message
ist	individual state signal.

4.6.2.3 Device handshake process

The HEF4738V controls the device's data traffic with two different handshake processes. These handshake processes tune the operation of the interface unit and the device during data transmission and reception. In both processes there are two signals involved: one generated by the interface unit, the other by the device.

Figure 4.31(a) shows the signals of the talker handshake process. The timing of the interface handshake process is also shown in the Figure. In the talker handshake process the data transmission is controlled by the dcd (don't change data) and the $\overline{\text{nba}}$ (new byte available) signals.

Figure 4.31(b) shows the timing of the listener handshake process. The signals participating in the handshake process are: the dvd (data to device valid), originating from the interface unit, and the rdy (ready for next byte), generated in the device.

If the data generation and transfer speed of the device are sufficiently fast then the two device handshake processes can be omitted.

4.6.3 Comparison of the MC68488 and the HEF4738V interface circuits

A common feature of the Motorola Type MC68488 and the Philips Type HEF4738V interface units is that they can be used, augmented with a few simple auxiliary circuits, for the interfacing of talker and listener devices of the IEC system. The external auxiliary circuits are required partly because it is not possible to achiev the 48 mA current load, specified by the IEC bus standard, with the single chip MOS-LSI circuits. Further auxiliary units are needed, because the 40 connections of the package are insuffiecint for the transfer of the complex input/output signals.

Despite the functional and manufacturing technology similarities, the structure and the operation of the two circuits are fundamentally different. The Type HEF4738V is an entirely hardware-realized IEC interfacing. The circuit is basically an LSI technology universal interface circuit, substituting for 40 to 50 traditional TTL circuits. The operational speed of an IEC device containing it will usually be determined by the data handling speed of the basic device. The interface unit allows about 200 kbytes/second maximum data transfer speed, appropriate to the 2 MHz clock signal that is controlling the input of the local messages. Its use does not present any limitations regarding the control of the interfaced device. It can be used for interfacing both traditional TTL circuits and microprocessor controlled devices.

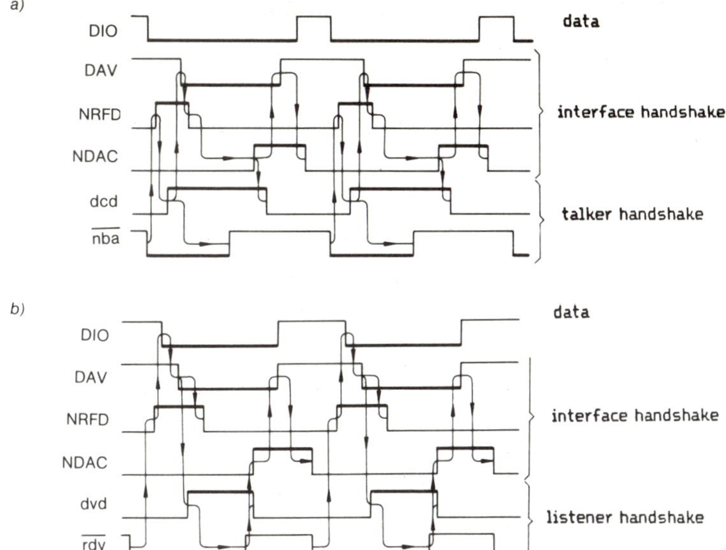

Figure 4.31 Handshake processes of the HEF4738V interface. (a) Talker handshake, (b) listener handshake

The Type MC68488 interface unit, on the other hand, can be used only with Type MC6800 or Type MC6802 microprocessors. This circuit is basically a software realized IEC interface and it is an I/O circuit whose operation is directed by the program controlling the microprocessor. As with all software-realizations, it is slower than a hardware unit handling identical tasks, its operational speed being about 10 kbytes/second in the normal (non-DMA) mode.

5 Design and construction of automatic measuring systems

The increasing interest in automatic measuring systems can be observed in almost every field of measuring technology. In itself this is a good thing, as these measuring systems are suitable for executing measuring tasks that are either mundane or can be solved only with great difficulty, if at all, using traditional measurement methods. However, measuring automation is an undertaking requiring considerable investment and careful consideration and therefore it is desirable to examine which aspects should be considered prior to the automation of a measuring process.

The user must examine three important aspects in the first instance:

(1) Does the automation of the measurement result in a worthwhile saving of manpower?
(2) Is the task to be executed one that cannot be solved without measuring automation?
(3) Is there a problem, so far avoided or ignored, that can be solved only by the automation of the system?

The answers to these questions, which cannot always be a simple 'yes' or 'no', are not sufficient to decide whether automation is the best solution in each case. However, new and important aspects, relevant to the whole investment, can come to light during the remaining stages of the assessment.

If it is clear from the above that the best solution to the problem is an automatic measuring system, then the selection of the measuring system may begin. This is a highly technical task; not only is a comprehensive knowledge of the measurement problems essential, but the responsible person must be familiar with the possible solutions and also be aware of the choices offered by instrument manufacturers.

Before the selection of the measuring system is discussed, let us examine the general requirements relating to an automatic measuring system.

(1) *Conformity*: The first and foremost requirement of an automatic measuring system is that it must satisfy the given measurement task and its operational conditions and environment. The purpose of the system must therefore be fully defined. Equally, the user must possess the necessary programming skills that will be demanded by the system or arrangements must be made for appropriate training.
(2) *Possibility of development*: In the present rapid development of measuring technology it is a most important requirement that the system should be suitable for later enhancement, when the tasks required of it change. The system design should therefore be carried out with this in mind. A modular structure with interchangeable units, or programmed control systems, where the program change ensures conformity to the new measuring task, are examples of how this may be achieved.
(3) *Economy*: A further basic requirement regarding automatic measuring systems is that of economy; however, economic returns are often difficult to predict. In practice, as well as saving in manpower, automatic systems provide less tangible advantages, such as the elimination of human error, greater accuracy, reproducibility and speed.

5.1 Design process of automatic measuring systems

In Figure 5.1 the entire cycle of the design process for an automated measuring system is shown. The first step is the realization of the system specification, considering the requirements and the realistic possibilities. The hardware and the software designers must collaborate in determining the necessary system characteristics for the detailed design.

Then follows the design of the hardware and the software, which ideally should take place simultaneously, on a constantly-interactive basis. The hardware designer must know the software in detail, as it will control the hardware designed by him. On the other hand, the software designer should also be

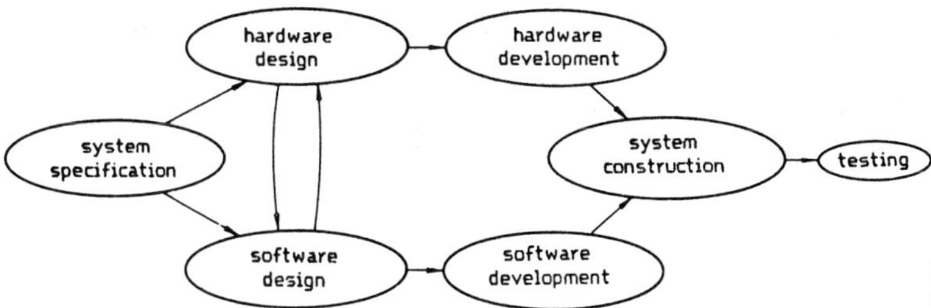

Figure 5.1 Design process of an automatic measuring system

aware of the possibilities and requirements of the hardware, as the system controlling software can be produced only by complying with these.

The interdependence of these two design stages is further emphasized by the fact that there are many functions in automatic measuring systems that can be realized by either hardware or software. This is especially true for universal interface systems, such as the previously discussed IEC interface. The optimal solution can only be reached by careful weighing of the advantages and disadvantages of the various realizations of the functions in question.

Very often the software realized solution is preferred in view of its inherent flexibility and due to reduction of the hardware costs. It may happen though, that the hardware realization solution must be chosen if, for instance, the speed requirements of the system demand it. In most cases a hybrid realization is the best solution; in other words, the hardware elements will deal with one part of the problem and the software with the rest.

On completion of the specification and system design phase, the parallel development of the hardware and the software elements can begin. Following this, after preliminary testing of the individual components, assembly of the system can proceed. In the standard IEC interface system the individual instruments are connected via standardized cables and connectors (Figure 5.2), which can be purchased commercially.

If there is a complex system, containing several instruments, it is not advisable to carry out the

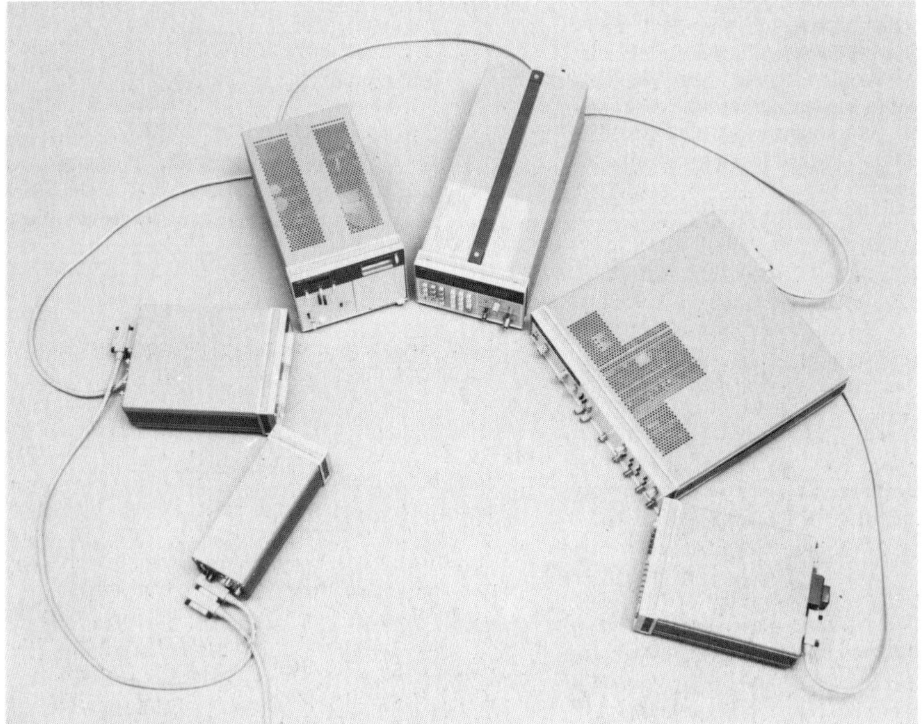

Figure 5.2 The connection of devices in the IEC system

assembly and the consequent testing in one step, as the number of potential errors is greatly magnified and their location is difficult to find. In such cases testing of the individual hardware and software subsystems is recommended, and when their operation is satisfactory, the assembly of the whole system can take place.

In the following – returning to the original theme of the book – the design of IEC-interface-based measuring systems is discussed.

5.2 Hardware design and development

The individual interface systems simplify the work of design engineers of automatic measuring systems, by eliminating problems of incompatibility. Thus the hardware design engineer can concentrate on the measurement process to be carried out and his task consists of choosing the most appropriate instruments and signal sources and ensuring their proper connection to the tasks and to each other.

5.2.1 Selection of measuring instruments and signal sources

There are two main aspects to consider when instruments and signal sources are selected for use in the IEC system. One is that the technical parameters (range, operational mode, etc) of the selected instrument should comply with the demands dictated by the measurement to be made. The other is that the instrument must have an IEC interface unit that allows the traffic of program and data transfer via the IEC bus.

The selection of the instruments according to their technical parameters will be dealt with first. This is a very important step in system design and very often the economic justification of the automatic system is determined by this. It is well known that the prices of instruments suitable for the same purpose can vary considerably, according to their sensitivity, accuracy and stability. The more expensive instrument is not always the better choice, especially if it means paying more for unnecessary features.

Graphical representation of the required parameters will facilitate the selection of the instruments. There are parameters to be measured (in the case of measuring instruments) and parameters to be generated (in the case of signal sources). In both cases the tolerance, required accuracy, etc, must be considered. In Figures 5.3 and 5.4 the necessary parameter plots are shown for the selection of a voltmeter.

In Figure 5.3 the required accuracy for measuring direct voltage is plotted as a function of the voltage. It is obvious from the graph that in our example the maximum accuracy (0.05%) is only required in a relatively narrow band, between 1 V and 10 V. This is generally the case in practice, i.e. the maximum accuracy or sensitivity of instruments is required in only a narrow band of their ranges.

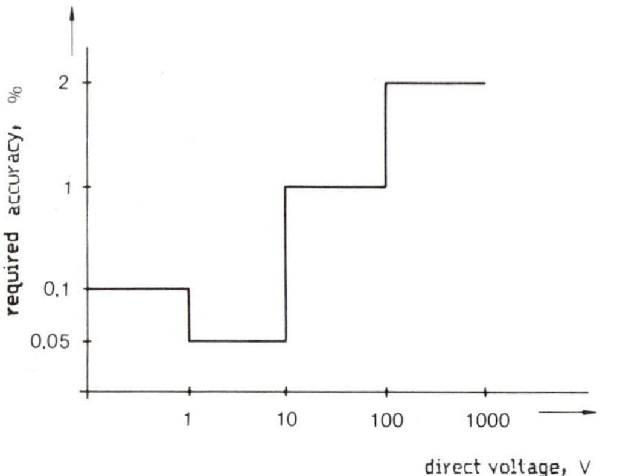

Figure 5.3 Parameter plot for direct voltage measurement

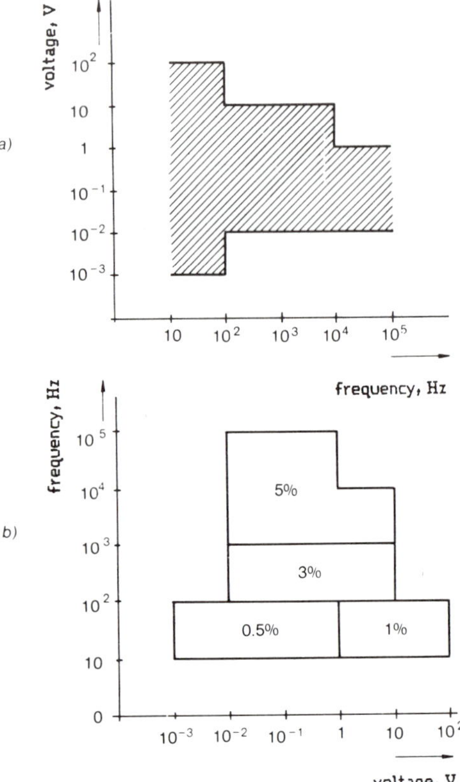

Figure 5.4 Parameter plot for alternating voltage measurement

INSTRUMENT	RESISTANCE Ω (10^{-6} 10^{-3} 10^{0} 10^{3} 10^{6} 10^{9}; μ m k M G)	CAPACITANCE F (10^{-15} 10^{-12} 10^{-9} 10^{-6} 10^{-3} 10^{0}; f p n μ m)	INDUCTANCE L MIN RESOLN	INDUCTANCE L MAX RANGE	D	Q	TEST FREQUENCY DC	Hz 60	120	400	1 k	10 k	1 M	BASIC ACCURACY %
DIGITAL LRC 296 (D)			0.001 μH	20,000 H	■	■			■		■			0.1
UNIVERSAL BRIDGE 293 (MB)			0.01 μH	1200 H	■	■	■				■			0.1 L, C 0.05 R
DIGITAL LRC 251 (D)			0.1 μH	200 H							■			0.25
UNIVERSAL BRIDGE 250DE (MB)			0.1 μH	1200 H	■	■	■				■			0.2 L, C 0.1 R
DIGITAL LRC 410 (D)			0.1 nH	2000 H	■	■							■	0.25
DIGITAL C METER 275 (D)					■						■			0.1
DIGITAL C METER 278 (D)					■				■					0.1
DIGITAL C METER 475 (D)					■						■			0.1
PRECISION C 701 (MB)					■			0 to 20 kHz						0.01, 100 Hz to 1 kHz
R COMPARATOR 262 (A)							■							0.01
R DEVIATION BRIDGE 263 (A)							■							0.02
R DEVIATION BRIDGE 506 (D)														0.005
LOW Ω OHMMETER 1700/1701B (D)														0.02
AUTORANGING 1700/1705 (D)														0.02
PRECISION RESISTANCE 242D (MB)							■							0.001
WHEATSTONE BRIDGE 231C (MB)							■							0.01

Figure 5.5 Tabulated characteristics of the R, L, C measuring devices. (Copy of the leaflet issued by the Ammerican ESI Company). Abbreviations: (A) analogue display, (D) digital display, (MB) manual balancing

In Figure 5.4(a) the required alternating voltage ranges are plotted as a function of frequency. In Figure 5.4(b) the required accuracy is shown as a function of both frequency and voltage.

Other properties to be measured, e.g. frequency, efficiency, etc, can be illustrated in a similar way. This graphic evaluation method can also be used for examining the parameters of signal sources.

The comparison of tabulated characteristics and ranges of instruments suitable for measuring several physical quantities (e.g. multimeters, R,L,C meters) can be cumbersome. The selection of the appropriate instrument is made easier if the physical quantities are indicated with vertical columns and the individual instrument ranges with horizontal bars.

Sometimes the instrument manufacturers themselves publish data in this form. Figure 5.5 shows the comparative features of Electro Sciences Industries Inc. (ESI) instruments. In the Figure, moving downwards, the parameters of R,L,C bridges, capacitance and resistance meters are shown. The Figure illustrates in a very clear pattern the ranges and the values of measuring frequencies and the error limits of the individual instruments. In a table prepared by the user the various makes of instruments would be listed, to aid in the selection of the most appropriate one.

The examples above cannot be used without modification. Our aim is to emphasize that the graphic representation method greatly simplifies the often very complex task of the system designer.

The examination of the analogue technical parameters is only part of the selection process for instruments to be used in the IEC system. Other, equally important aspects of the selection are concerned with selection of the system characteristics, programming possibilities and the form of the output. In considering these aspects it should be ascertained that all the relevant operational modes and ranges of the instrument can be set up from the interface; in other words, the format and the timing of data transferred from the instrument to the bus must comply with the system requirements. The instrument catalogues and data sheets do not always contain detailed information on program control and in such cases it is worthwhile to request a programming card from the manufacturer. Such cards contain the program control in detail and may even list some simple programs.

The programming of an instrument and the output of data is closely related to the intrinsic intelligence of the instrument. Microprocessor-controlled instruments are easier to program and are capable of producing data output in arbitrary format. The advantages arising from this are discussed in the section dealing with the programming of IEC systems.

5.2.2 Selection of the control unit

It is without doubt that one of the most important characteristics of an automatic measuring system, the accuracy of measurements, is determined by the parameters of the instruments and signal sources used in the system. However, this is only one part of the factors affecting system performance. The operational speed and easy handling of the system are equally important factors and these are closely related to the characteristics of the control unit.

A simple and clean-cut formula for the selection of the controller cannot be given, as every automatic system must be considered on its own merits. The most important aspects are:

- *Operational speed*: How fast is the controller with respect to the IEC interface? Is there a DMA facility? (Remember that desk calculators do not handle DMA or accept interrupts.)
- *Capacity of main memory*: Is the memory capacity sufficient for storing the program and the measurement data? The program storage of machine code will be more efficient than an interpretive language and the user will need to know how the machine stores its data.
- *Program language*: Can the controller be programmed in BASIC, the most frequently used language in engineering practice or in some other commonly used high level language? Are there any specific IEC bus control commands in the instruction set? Does the user have programming experience in the given language?
- *Availability of peripherals*: Are the built-in peripherals of the controller suitable for the given task? If there should be a need for external peripherals, how readily can they be interfaced to the controller?

In traditional automatic measuring systems using non-standard interfaces, there is data traffic only between the controller and the devices. Hence the control unit and the rest of the system are strictly inter-related and the arrangement cannot be altered at all, or only by the complete redesigning of the controller's interface.

In the IEC interface system the controller can be selected independently from the instruments and the system configuration. If the configuration of the system is altered, or even if the controller is changed (e.g. use of a computer instead of a calculator), the interface units need not be redesigned. Another important difference is that in the IEC systems it is not always the controller's task to program-control the other devices and to process data. According to definition, the IEC system control directs the system operation and data traffic. The selection of the controller must be made according to the system's tasks and required automation level. To illustrate this, a few examples will be given.

In the simplest IEC system a talker and a listener device can be connected, omitting the controller (Figure 5.6). In this configuration the talker device (voltmeter) sends the measurement data in the 'talk only' operational mode (via the IEC bus) to the listener device (printer) that is operating in the 'listen only' operational mode. This means that the data transferred to the bus by the talker is accepted by the listener automatically, without any additional addressing, the instruments being set up manually. As there is no controller, the repetition time of the measurements is determined by the internal timing of the slowest device.

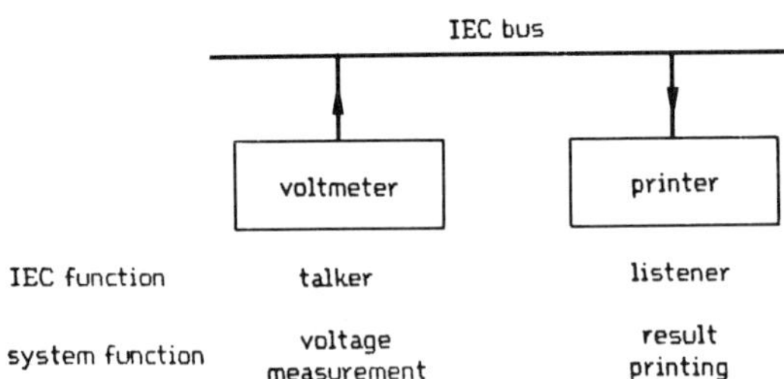

Figure 5.6 Simple IEC system, containing talker and listener devices

If a system contains three or more devices, some form of control is necessary. The simplest form of an IEC controller is a sequence controller that addresses the talkers one-by-one, in a predetermined order and timing. The talkers sequentially transmit the measurement data to the printer (Figure 5.7). As in the previous example, the instrument-programming data are not transmitted to the bus, the instruments being set manually.

If the instruments of the system are to be programmed from the interface, then a fixed program controller containing a ROM or a magnetic disk or tape reader must be used (Figure 5.8). In this configuration the controller can handle programming tasks, as well as direct the data traffic by allocating the listeners and talkers. Owing to the nature of the control, the structure of the system and the measurement sequence cannot be altered.

The intelligent controller (e.g. programmable calculator, mini- or microcomputer) incorporated in the most advanced forms of the IEC systems,

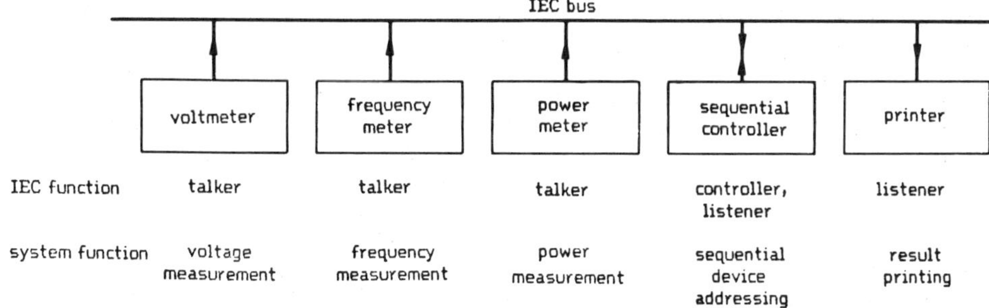

Figure 5.7 IEC measuring system containing a sequential controller

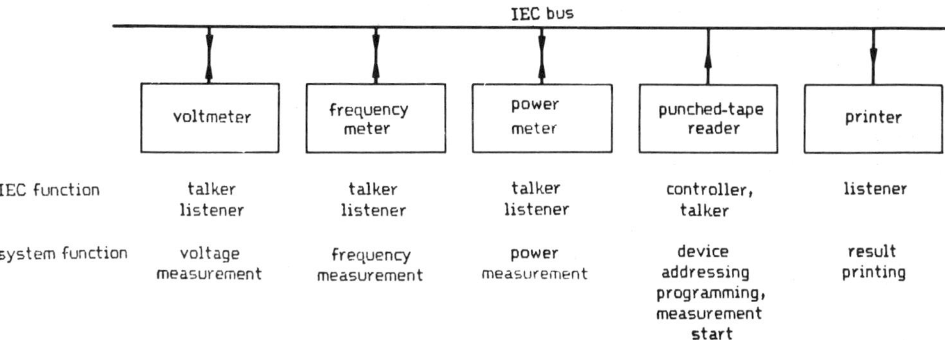

Figure 5.8 Structure of an IEC system containing a fixed program controller

already has automatic data processing and decision making features (Figure 5.9). Furthermore, these measuring systems allow interactive intervention with the user.

The above examples show that the nature of the measurement task defines the character of the controller. The advantages of the simpler systems using sequence controllers are that they are cheaper and a knowledge of computing techniques is not necessary. Their disadvantage is the rigidity of both the structure and the operation of the system.

The choice between programmable calculators and computers is not clear cut. The use of calculators is preferable for controlling simple or medium complex systems, as the program writing and construction is simpler but allowance must be made for their inability to accept interrupts or DMA. On the other hand, computers are better for controlling complex systems that may require large memory capacity and fast operational speed.

5.2.2.1 *Interactive control*

In Chapter 3 dealing with control units, the devices used for controlling measuring systems of various automation levels were discussed in detail. These devices can be classified in three categories according to the programming mode:

(a) programmable in assembler language,
(b) programmable in a compiler language,
(c) programmable in an interpretive language.

The structures of the above three categories are shown in Figure 5.10.

The programming of assembler language controllers is done in an individual, machine-specific language. The program given in a mnemonic code is translated by the assembler into the machine code that is intelligible to the hardware. In the on-line operational mode the user can only communicate with the controller in machine code (Figure 5.10(a)).

The programs of high-level language controllers are translated by the compiler or interpreter into machine code. In the former the compiler translates the entire program at once and on-line interaction between the user and the controller is again only achieved in machine code (Figure 5.10(b)).

In contrast to these approaches, the interpretive language programmable controllers ensure that the user can interact in the on-line operational mode using a higher-level language (Figure 5.10(c)). This

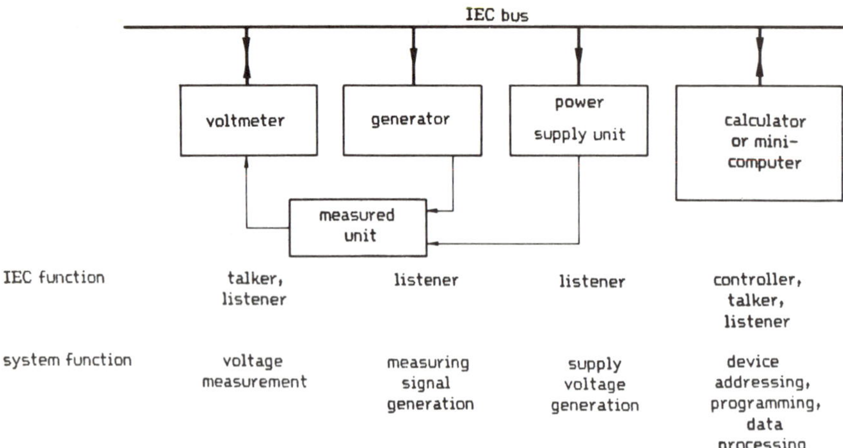

Figure 5.9 IEC system containing an intelligent controller

is due to the fact that the interpreter translates, interprets and tests the input program lines one-by-one. This system has disadvantages, in that the interpreter is much slower and requires larger memory capacity than the compiler. However, the high-level language on-line interaction facility is such a great advantage that these systems are used for automatic measuring system controller almost without exception.

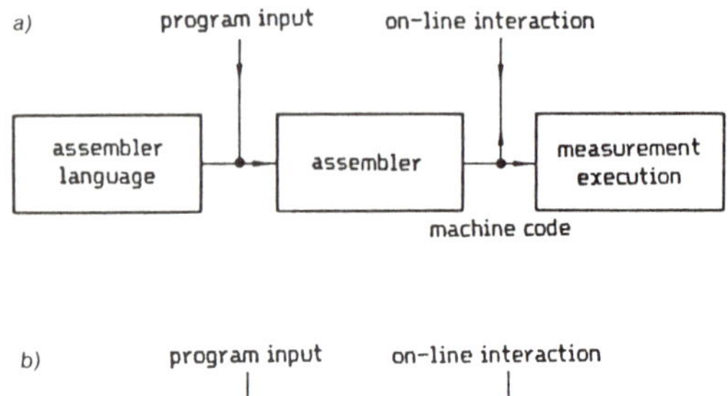

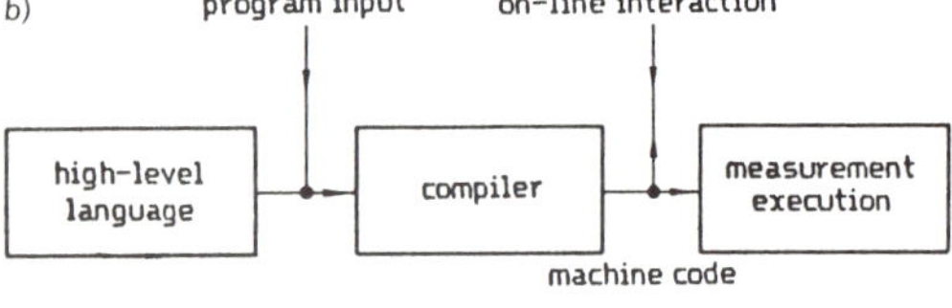

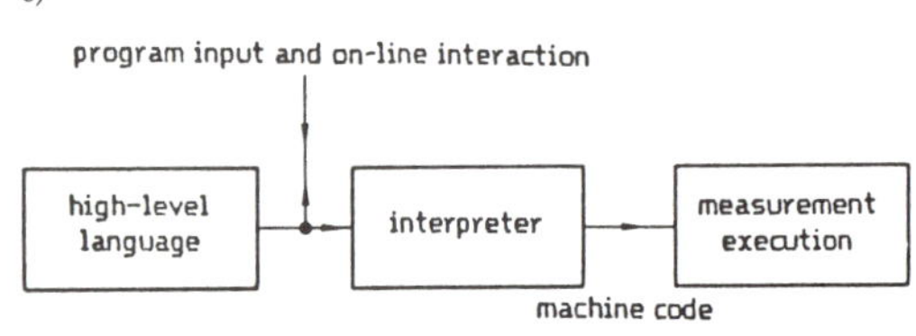

Figure 5.10 Program input and on-line interaction for various programming languages. (a) Assembler, (b) compiler, (c) interpreter

5.2.2.2 The efficiency of the program

In general terms, the efficiency of a computer program means that the given task is solved in the shortest possible time, using the least amount of memory capacity. The time requirements of the program and its control responses seldom present problems in the programming of automatic measuring systems. The operational speed of an automatic measuring system is limited by the course of the tested physical processes and the response time of the devices used in the system. In consequence, the operation of the controller must often be slowed down by incorporating delays.

The other program efficiency factor, the economical use of the memory, is of great importance in measuring technology as well. In IEC systems the memory capacity requirement of a program is determined mainly by the program data requirements of the instruments and the amount and format of measurement data.

Microprocessor controlled, intelligent instruments are capable of giving the measurement data in the required format, as a result of a few simple program commands. Thus the memory capacity requirement of the program is decreased and program writing becomes simpler.

5.3 Software design and development

The operation of an automatic measuring system is controlled by a program operating through the hardware part of the control unit. The program will determine the sort of measurements to be executed and their timing, the format and timing of the measurement data supplied by the devices and the kind of operations carried out by the controller on these data prior to graphic or alphanumeric display.

The cost of program development can be a major part of the cost of the whole system. Hence, any facility that reduces programming time is just as important as the selection of the appropriate instruments during the hardware design phase of the systems.

5.3.1 The construction of the user's program

In the IEC system all user's programs consist of two parts. The basic utilities module containing the routines (subroutines, procedures, library modules) relating to the programming and data traffic of the individual devices is one of these. The construction of the utility program segments is carried out with the aid of machine handbooks or code tables containing the detailed programming data of the instruments.

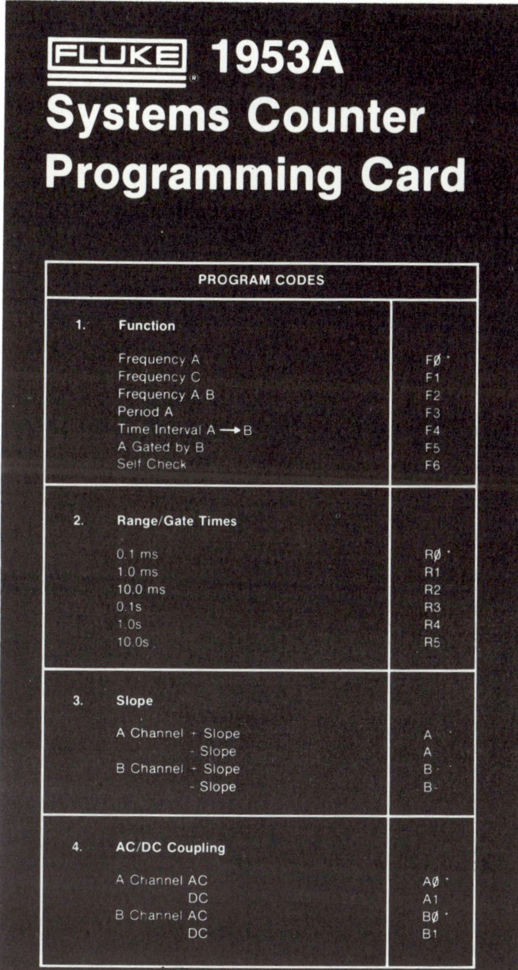

FLUKE 1953A
Systems Counter
Programming Card

PROGRAM CODES	
1. Function	
Frequency A	FØ
Frequency C	F1
Frequency A B	F2
Period A	F3
Time Interval A ⟶ B	F4
A Gated by B	F5
Self Check	F6
2. Range/Gate Times	
0.1 ms	RØ
1.0 ms	R1
10.0 ms	R2
0.1s	R3
1.0s	R4
10.0s	R5
3. Slope	
A Channel + Slope	A
- Slope	A
B Channel + Slope	B
- Slope	B-
4. AC/DC Coupling	
A Channel AC	AØ
DC	A1
B Channel AC	BØ
DC	B1

Figure 5.11 Code table of an IEC-compatible instrument (Copy of a leaflet issued by Fluke)

Figure 5.11 shows details from the code table of an IEC compatible instrument, the Fluke Type 1953A frequency meter. The code table detail, shown in the Figure, contains the codes for the instrument's operational mode (function), the range/gate times, the slope and the coupling mode (AC/DC). The Table also gives the data formats of the measurement data outputs.

The IEC standard, complemented by the code and format recommendation, makes the construction of the utility program much easier. Furthermore, the marketing companies of automatic measuring systems usually provide customers with program routines executing the programming of the instruments and the most frequently used measurement cycles. These routines are developed for certain instruments and can be used only with a particular controller. For example, the high-frequency instrument manufacturing company of Rohde & Schwarz give the basic routines of their Type SPMU radio test assembly appropriate to the Tektronix Type 4051 graphic calculator. This measuring system, shown in Figure 5.12, is suitable for the automatic testing of VHF transmitter/receiver characteristics. The utility program, stored on magnetic cassette, is loaded into the memory of the calculator; the user can then start and control complex measuring cycles with simple instructions.

The construction and use of the utility module is related to the automation level of the measuring system and to the nature of the measurement task. For instance, there is no utility module in sequence controller directed systems, as the user sets the instruments manually. If often happens that the most important element of the utility program is not the instrument programming section, but the data-processing routines (e.g. linearising or averaging routines).

The aim in writing utility programs is to ensure that the sections handling separable tasks should be self-contained units within the program. The division of the program into such segments will:

- make program writing easier,
- simplify program debugging and corrections, and
- facilitate program development.

The first step in the utility program construction is the drawing of a flow diagram (Figure 5.13). In working out the diagram, the steps to be executed must be arranged in a logical sequence according to the tasks and taking into account all the possibilities

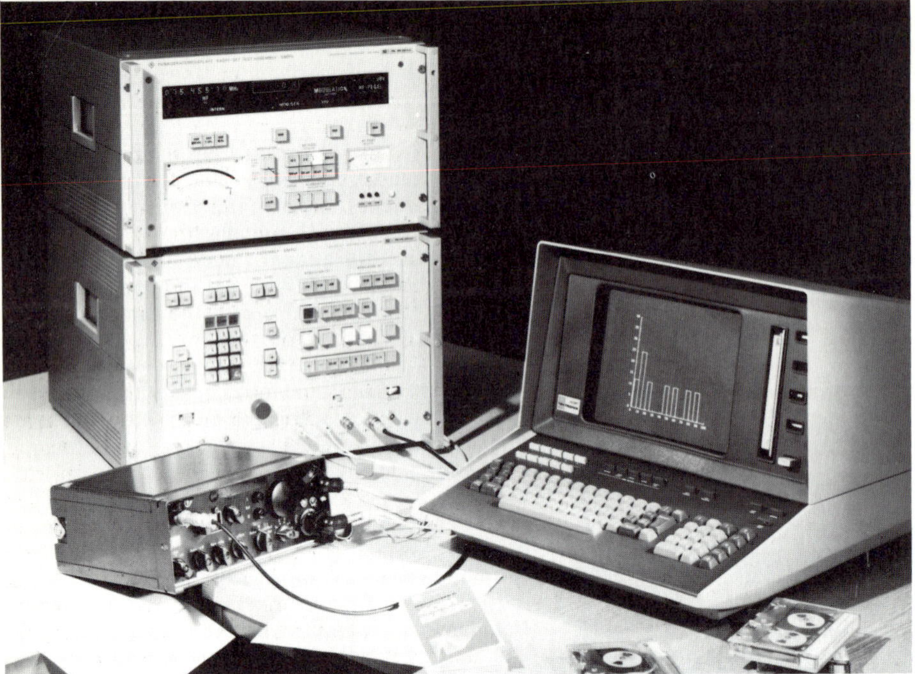

Figure 5.12 Tektronix 4051 graphic-calculator-controlled automatic transmitter/receiver measuring system

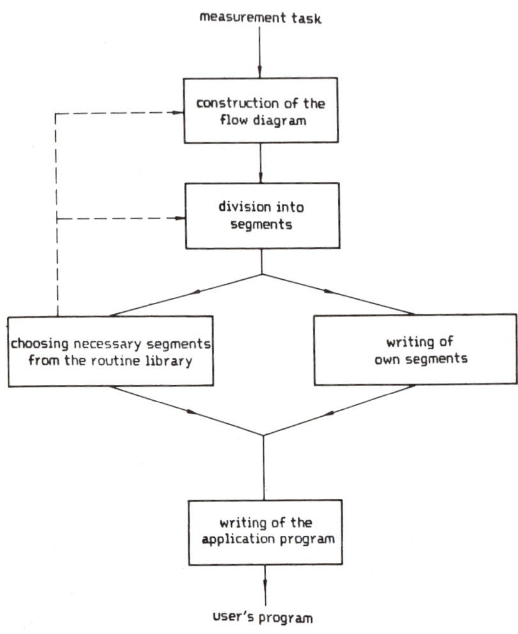

Figure 5.13 Flow diagram of the user's program construction for operating an automatic measuring system

provided by the system. This work requires considerable experience, as well as a thorough knowledge of instrumentation and of the controller's operation.

The program construction is continued with the segmenting and existing segments are selected from the routine library. This work can influence the segmenting and even the production of the flow diagram. This is referred to by the broken lines in the Figure.

The next step is writing of one's own program segments, followed by the writing of the appropriate application program calling the various segments.

5.3.1.1 *Application programs*

The other part of the user's program is the application (job) program, describing the system operational process and the algorithm of the measuring cycle. The application program determines the event starting the measuring process, the sequence of the measurement cycles, the number of measurements within a measurement cycle and the destination of the measurement data.

Before the application program can be written, the address allocation must be made. Table 5.1 lists the listener and talker addresses applicable in the IEC system according to the setting of the address

Table 5.1 IEC instrument addresses

Address switches 12345	*Listener address*	*Talker address*
00000	SP	@
10000	!	A
01000	"	B
11000	#	C
00100	§	D
10100	%	E
01100	&	F
11100	'	G
00010	(	H
10010	)	I
01010	*	J
11010	+	K
00110	,	L
10110	–	M
01110	.	N
11110	/	O
00001	0	P
10001	1	Q
01001	2	R
11001	3	S
00101	4	T
10101	5	U
01101	6	V
11101	7	W
00011	8	X
10011	9	Y
01011	:	Z
11011	;	[
00111	<	\
10111	=	]
01111	>	^

switches contained in the devices. With the appropriate setting of these two switches listener and talker addresses are allocated simultaneously. There are some fixed, or pre-set, address devices, in which case it is advisable to keep those addresses and use the remaining unassigned ones for the other devices of the system.

In any one system there cannot be two identical talker addresses. Furthermore listener addresses can be identical only if the devices are to receive exactly the same information. It is advisable to compile the addresses of the system's devices in a table, as it will be needed continuously during program development.

After the address allocation the writing of the application program may begin, using the flow diagram and the existing utility program. The process of writing an application program will be demonstrated in the next section, using a practical example; accordingly, only a few selected aspects are emphasized in the following. In the practical example and in the following discussion the most commonly-encountered BASIC language programs will be used.

In the application program the individual instrument programming and measuring operations follow a predetermined sequence. Generally, the first task is the setting of the supply voltage and the output signals of the signal sources. The setting of the measuring instruments then follows, after which the actual measurements are made. To design the appropriate sequence and timing of the operations, the programmer must know the execution time of the operations present in the program and the setting-up time of the instruments, etc.

Additionally, factors that are automatically achieved during manually-controlled measurement should also be considered during the writing of the application program. For example, when measuring noise levels with an analogue voltmeter, an average value is measured due to the inertia of the pointer. The fast digital voltmeters used in automatic measuring systems measure the momentary value of the noise voltage in a single measuring cycle. The average value is obtained only if we instruct the voltmeter to take several samples and average them. Thus there are several program steps and an arithmetic operation that will be necessary for a seemingly simple measurement.

Switching transients present a similar problem in automatic measuring systems. If the frequency or the level of the output signals of the generators, which control the inputs of the measured units, are changed, a period of time must elapse before equilibrium of the measured characteristic is attained. This problem is not present during manual measurements, as there is at least a few tenths of a second lag between the setting of the generator and the reading of the instrument, during which these transients die away. In automatic measuring systems it is advisable to ensure the appropriate delay of the instrument, as in the following example:

```
60 WAIT 1000
```

This line of program can be inserted in the program at any place; for instance, it means a 1-second delay in the BASIC program language of the Type HP9830A calculator.

Subroutines that are repeated many times in the measurement program can be called simply with the program line number. Figure 5.14 shows an example of this. In program step 990 the execution is transferred to the program line 670, as a result of the GOSUB command. The subroutine is terminated with a RETURN command, then the execution goes back to the instruction following the subroutine call.

In programming automatic measuring systems the most frequently used commands are unconditional and conditional jumps.

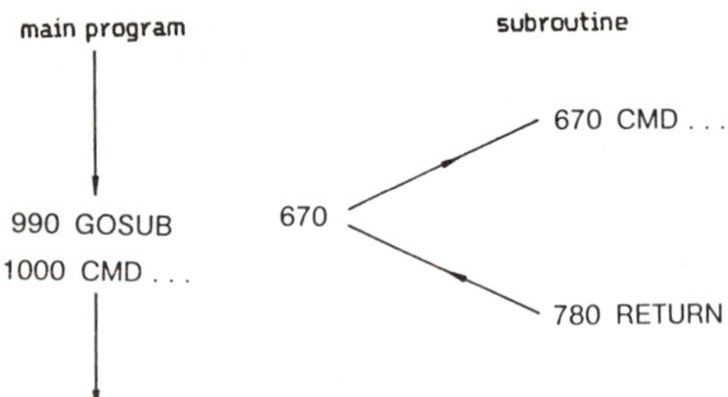

Figure 5.14 Inserting a subroutine into a BASIC language program

The unconditional jump instruction is assigned by the GOTO command, giving the line number of the instruction from which the program execution should continue.

Let us examine the following program in detail:

```
10 LET X = 2 * Y
20 GOTO 40
30 LET X = X ↑ 2
40 PRINT X
```

In the above program after the value allocating line 10, the next step is the printing of the X value, (PRINT X), as a result of the 20 GOTO 40 instruction.

The execution of the conditional jump instruction given in the IF ... THEN format, depends upon whether a mathematical or logic condition is met. If the condition is met (TRUE), then the program continues with the program line following the THEN expression. If the condition is not met (FALSE), then the program continues with the next line. If the condition was set up in such a way that the no-jump continuation of the program is appropriate for the FALSE answer, additional GOTO instructions will not be necessary, for example:

```
70 IF A < = B THEN 100
80 PRINT "A IS GREATER THAN B"
90 GOTO 70
100 PRINT "A IS SMALLER THAN OR
    EQUAL TO B"
```

Finally, we would like to draw attention to a programming possibility that can make the operation of semi-automatic measuring systems much easier. In constructing conversational or dialogue type programs, messages can be incorporated that instruct the operator to provide manual intervention. During manual operations the program execution is interrupted and can be resumed only by manual input by the operator. Documentation on the program is important and it is vital to comment the source program fully.

5.3.2 A practical example for constructing a user's program

The measuring system in our example can be used for determining the electrical length or the permittivity of high-frequency cables to high accuracy. In the measurement the principle is used that a voltage impulse applied to the input of an open-ended high-frequency cable travels the length of the cable and is reflected from the open end in the same phase, and that it reappears at the input, having a smaller amplitude due to the losses suffered.

The time difference between the positive going slopes of the original impulse and the reflected impulse is equal to twice the value of the cable delay time and is dependent on the propagation speed of the signal and on the length of the cable.

The propagation speed (v) of an electric signal in a given signal cable is determined by the magnetic permeability (μ) and permittivity (ε) of the surrounding medium thus:

$$v = \frac{1}{\sqrt{\mu\varepsilon}}$$

By rearranging the above universal equation, introducing the condition $\mu_r = 1$, the cable delay time (t) will be given by:

$$t = \frac{1\sqrt{\varepsilon_r}}{3 \times 10^{10}} \text{ seconds}$$

where ε_r is the relative permittivity of the cable insulation and 1 is the length of the cable in cm units.

It is clear from the formulae that by measuring the delay time for a given permittivity the cable length can be determined; alternatively, the value of relative permittivity can be obtained for a given cable length.

The measuring arrangement is shown in Figure 5.15(a). The signal of the impulse generator is transmitted through an interfaced connector to the common input (AB) of the counter executing the time measurement, and to the measured and reference cables via a high-frequency switching unit. The measurement is made by comparison, the high-frequency switching unit connecting first the reference cable to the measuring system, followed by the unknown cable. For the time measurement the starting levels of the counters A and B inputs must be set to different values (Figure 5.15(b)). The τ time difference can be determined from the two measurements, then the appropriate characteristics can be calculated from the formulae given above.

The Type HP9830 calculator controlling the measuring system is interfaced to the system units via the IEC bus. The address allocation is contained in Table 5.2. It is advisable to set the addresses of

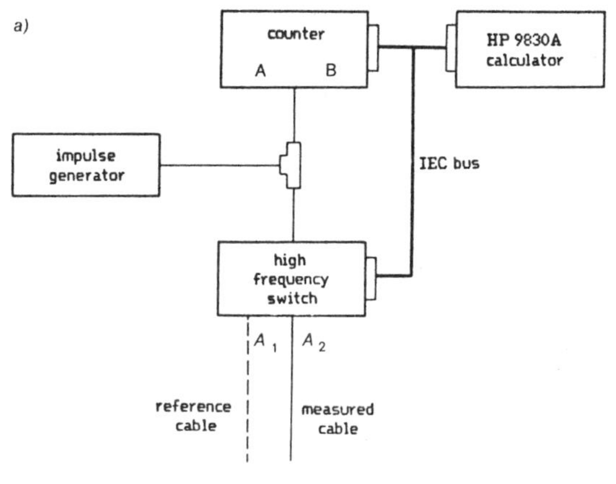

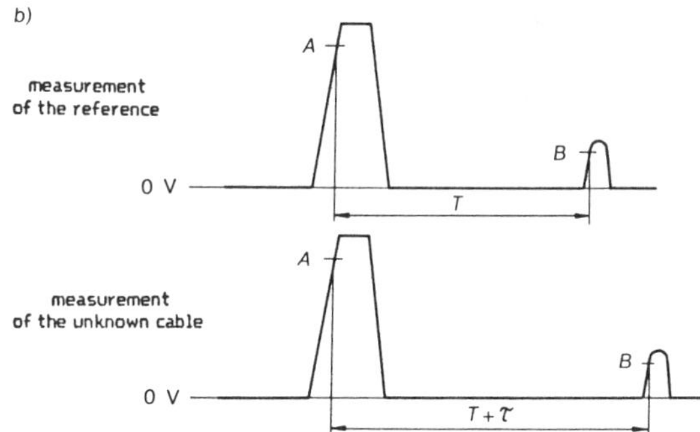

Figure 5.15 Automatic measuring system for determining the electrical length of permittivity of a high-frequency cable. (a) Measuring system structure, (b) measured signals appearing on the counter input

the devices before the interface cables are attached, because in many devices the cable deliberately obscures the address switches so as to exclude the possibility of accidental switching, which would disturb the operation of the system.

It is not necessary to control the impulse generator with the calculator, but its range of time measurement must be selected according to the length of the measured cable.

The flow diagram of the measurement is shown in Figure 5.16. In the flow diagram the appropriate line numbers of the BASIC language program are also shown. In constructing the program, our aim was to ensure that it should be equally suitable for determining both the cable length and the permittivity value. For this reason the conversational style program contains two conditional and several unconditional jump instructions.

```
 10 DIM B $ [3]
 20 DISP "LENGTH MEASUREMENT"
 30 WAIT 1000
 40 DISP "LENGTH MEASUREMENT - YES
    OR NO";
 50 INPUT B $
 60 IF B $ [1,1] ="Y" THEN 120
 70 DISP "REL PERMITTIVITY MEASURE-
    MENT"
 80 WAIT 1000
 90 DISP "LENGTH (CM)";
100 INPUT B
110 GOTO 140
120 DISP "REL PERMITTIVITY";
130 INPUT C
140 CMD "?U>","A1"
150 WAIT 50
160 CMD "?U>","12F3G>E89:I1","?J"
170 WAIT 50
180 CMD "?U*","J1","?J5"
190 ENTER (13,*)Y
200 CMD"?U>","A2"
210 WAIT 50
220 CMD "?U*","J1","?J5"
230 ENTER (13,*)X
240 Z=(X-Y)/2
250 IF B $[1,1]="Y" THEN 290
260 FIXED 3
270 PRINT "REL PERMITTIVITY", (Z*3E+10/
    B)↑2
280 GOTO 10
290 FLOAT 3
300 PRINT "DELAY TIME (SEC) =",Z
310 FIXED 2
320 PRINT "LENGTH (CM) =",Z*3E↑10/
    SQR(C)
330 PRINT
340 END
```

Table 5.2 Address allocation of the measuring system

Instrument	Talker address	Listener address	Address switches				
			A5	A4	A3	A2	A1
Calculator	U	5	internally set				
Counter	J	*	0	1	0	1	0
High frequency switch	Λ	>	1	1	1	1	0

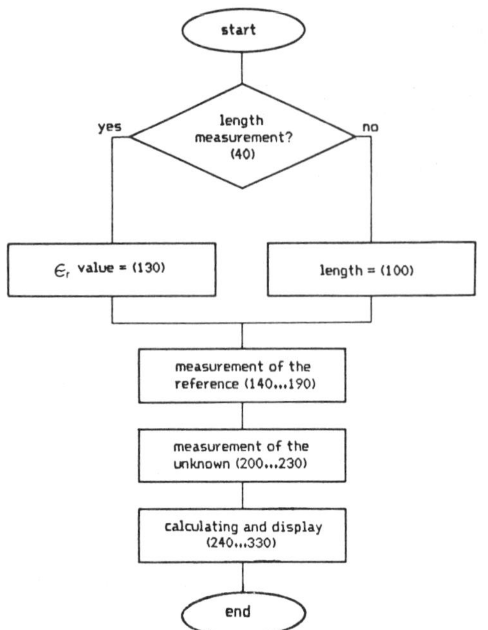

Figure 5.16 Flow diagram of the automatic measurement of a cable

The functions of the individual program lines are:

10 Dimensioning of the B string (maximum 3 characters).

20,30 Display of program title for 1 second.

40 Display of the conditional jump condition. As a result of the ; character, the display is maintained until the B string has been put in.

50 Input of the value of the B string (YES or NO).

60 Conditional jump instruction; the condition is, does the B string begin with Y? If 'yes', the execution is transferred to program line 120. If 'no', the program continues with the next line.

70,80 Display of 'relative permittivity measurement' for 1 second.

90 Asking for the cable length value, the display is maintained until B variable is put in.

100 Input of variable B.

110 Unconditional jump to program line 140.

120 Asking for the value of relative permittivity (variable C).

130 Input of variable C.

140,150 IEC controller program command for setting the high-frequency switch (reference measurement).

160,170 IEC controller program command for setting the counter.

180,190 Requesting the measurement data from the counter (variable Y).

200,210 Setting the high-frequency switch to measure the unknown cable.

220,230 Requesting the measurement data from the counter (variable X).

240 Arithmetic expression for the calculation of delay time.

250 Conditional jump instruction; the condition is, does the B string begin with Y? If 'yes', the execution is transferred to line 290. If 'no', the program continues to the next line.

260,270 Instruction to provide the required display format (fixed decimal point, 3 digits) and arithmetic expression for calculating and printing the value of relative permittivity.

280 Unconditional jump instruction to program line 10.

290,300 Instruction to provide the required display format (floating decimal point, 3 digits) for printing the value of the delay time.

310,320 Instruction to provide the required display format (fixed decimal point, 3 digits) and arithmetic expression for the calculation and printing of the cable length.

330,340 Extra line feed, end of program.

6 Instruments and measurement methods for functional testing of IEC systems

A feature of IEC measuring systems is that it is not usually the task of the manufacturer to assemble the complete system. This will mean that the unavoidable functional errors and problems experienced during set up must often be eliminated by the user.

Such errors and problems influencing the operation of IEC measuring systems can be of either software or hardware origin. A software error can be a simple coding or data error, or perhaps misunderstanding of the mode of operation of programmed control instruments. Hardware errors can be caused by internal faults in the instruments or defects in interconnection. In many cases problems are caused by a combination of both sources of error. For instance, a typical hardware error may be low insulation resistance at the IEC connector of a device causing an error in range setting. This, however, is a phenomenon which may also be created by a software error.

Due to the bus structure of the IEC interface it can be difficult to locate the device transmitting erroneous data or commands to the bus and special instruments have been devised to facilitate testing of data in IEC systems. These began to be available around the same time that the IEC system gained universal acceptance. At present there are two major trends in such instrument development. These are the use of general purpose instruments for testing programmed control digital circuits, the so-called 'logic analyser', suitable for a variety of bus systems and the use of instruments dedicated to IEC system testing.

In the following sections, which major on IEC applications when the instruments are described, comparisons will be made to the ubiquitous oscilloscope, so commonly used for analogue measurements. It is assumed that the reader is familiar with the operating principles of this instrument and is aware of its possibilities.

6.1 Requirements concerning instruments

Before IEC system analysers are discussed in detail, let us summarize the basic requirements concerning these instruments.

(1) *They should permit simultaneous testing of logic levels on at least 16 signal lines*

In the IEC interface system the individual devices are connected via a 16-line cable. Each state of the interface is determined by the combined logic level of the 16 signal lines and hence their simultaneous parallel testing is essential.

(2) *They should be able to store the measured samples*

One of the features of digital technology is that the majority of signals present in circuits are aperiodic. This means that the oscilloscope, suitable for testing periodic signals, cannot be conveniently used for testing digital circuits.

Logic state analysers designed for the functional testing of digital circuits have their own memory units. They take samples of the tested signal, store it in their memory in a sequential format, then convert it into an analogue voltage value when it is displayed on the cathode-ray tube. More recent analysers can be instructed to record only data of particular interest or to undertake real-time data processing prior to display so as to facilitate the interpretation of results.

(3) *They should be equally suitable for synchronous and asynchronous measurements*

In digital measurement technology, 'synchronous measurement' means that in the input unit of the instrument the sampling is controlled by a signal originating from the measured circuit. This signal is not necessarily the clock signal but can be any other signal that participates in the timing of the operation of the tested circuit. If the external synchronizing signal is chosen suitably, the instrument will evaluate the signals sent to its input in exactly the same way as the tested circuit does.

In the case of 'asynchronous measurements' the operation of the instrument is controlled by an internal clock signal and the sampling input is independent of the operation of the measured circuit. The advantage of asynchronous measurement is that data may be sampled at a faster rate, allowing narrow transient impulses (e.g. hazard signals, noise spikes, etc) to be detected, using a suitably high internal clock frequency. This is not possible with synchronous measurements.

(4) *They should contain a combinative trigger circuit*

During the functional testing of digital circuits the recognition of a serial or parallel (according to the data transfer method) bit-pattern or word is necessary for the triggering of the instruments. This method is called 'combinative triggering' and allows several types of word recognition to be undertaken: simple, sequential, nested and non-sequential.

Simple word recognition permits a word, defining a single event, to be recognized, whilst sequential word recognition allows a sequence of simple events to be identified.

Nested word recognition is the ability to monitor conditional branching.

Non-sequential word recognition is similar to the first two cases, except that the trigger can follow a decision-dependent algorithm.

(5) *They should allow digital delay (according to the number of events)*

The delay facility present in most oscilloscopes shifts the triggering of display by an arbitrary pre-set time interval. In digital circuits, delay can be introduced according to the number of events, i.e. event counting.

(6) *They should have a variety of display formats*

The three basic requirements for display are: timing diagram, state table and mnemonic disassembly.

The timing diagram is particularly useful in locating hardware faults, whilst the state table allows the state of the circuit under test to be observed in tabular form. Mnemonic disassembly displays the data in the state table in a more readily understandable form.

As an example, let us assume that we would like to test the IEC bus data traffic with our instrument. If a software error is suspected and we wish to compare the ASCII characters transferred on the DIO1...DIO8 signal lines with the program data, then it is better if we see the character G on the screen rather than the corresponding 1110001 bit-pattern. However, if we need to investigate a hardware error, then a binary display will also be essential.

(7) *They should be suitable for active participation in IEC systems*

The operational speed of an IEC system is determined by the slowest device participating in the data transfer. If our instrument is capable of active participation in the operation of the system, then the data transfer can be slowed down by it and certain fault conditions can be simulated, etc.

There are two categories of IEC system analysers meeting these operational requirements:

(a) logic state analysers with IEC auxilliary units, and
(b) special IEC system test instruments.

The logic state analysers equipped with IEC auxiliary units are passive instruments. Owing to their high input resistance, they do not load the tested system and thus do not influence its operation. In contrast to this, the special IEC system test instruments are generally active devices that load the system as units do and participate in its operation.

6.2 Logic state analysers with IEC auxiliary units

At present logic state analysers provide the most versatile means of functional testing of digital circuits. As has already been stated, although these devices resemble oscilloscopes in many aspects, there are fundamental differences between them, both in operation and structure. The most important of these is that logic state analysers separate the data collection and display in time and frequency by digital storage. They are therefore more precisely digital domain analysers. The data collection is carried out at a speed identical to the operational speed of the tested circuit and the display speed is chosen to suit the operator and to allow ample time for evaluation.

Figure 6.1 shows the simplified block diagram of a logic state analyser. In the input unit the amplitudes of the incoming signals are measured against arbitrarily-set threshold limits by comparators. The sampling and the conversion to binary format also take place here, controlled by an external or internal clock signal. The former permits synchronous and

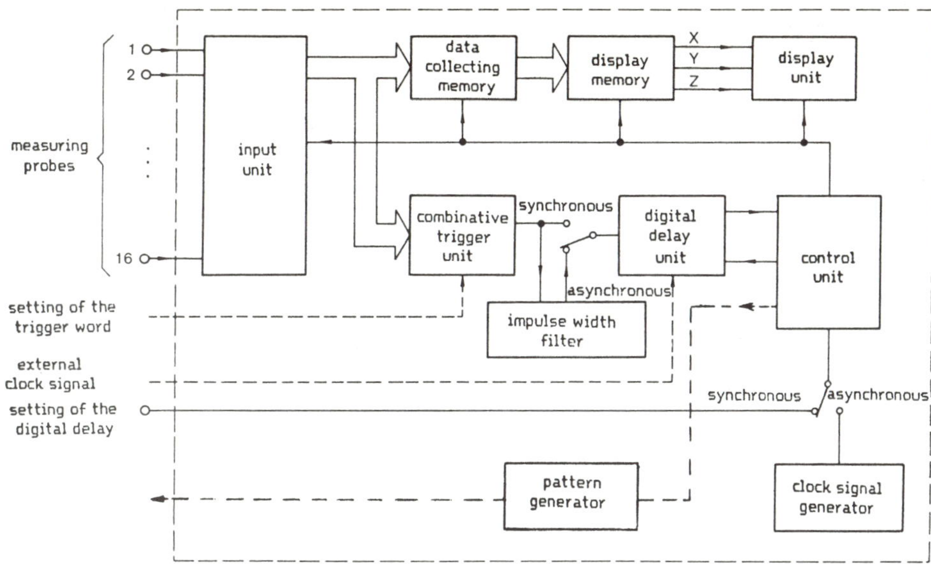

Figure 6.1 Structure of logic state analysers

the latter, asynchronous measurements to be made. Some analysers also include a pattern generator, shown with dotted lines in the Figure. The pattern generator allows the device under test to be stimulated from the analyser. The generator can be programmed with the most appropriate data for the device including routines which are known to be satisfactory.

The binary samples taken from the measured signals are transferred from the input unit to the data collection memory. One input per clock signal requires one bit of memory capacity. The memory stores the data arriving from the input by stepping them sequentially whilst filling up.

The incoming data is monitored by the combinative trigger circuit, giving the trigger signal when the required word appears on the input. Selection of the trigger word is carried out with the three-state (1,0,X) selector switches found on the front panel of the logic analyser. In asynchronous operational mode the selected trigger word can appear for a short time on the input, causing a false trigger. This is due to the time lag of data arriving on the parallel channels and is prevented by a built-in pulse duration filter.

The output signal of the combinative trigger unit is transferred to the digital delay unit. The value of the delay is set by switches located on the front panel of the logic analyser and the analyser is triggered only after the set delay has passed.

After the filling up of the data collecting memory, the data are transferred to the display memory by the control unit and the generation of the information appearing on the display is executed from here. The frequency of the reading is independent of the clock frequency used for filling up the data collecting memory. It is necessary to have a display on the analyser screen that can be evaluated easily; hence the stored data is refreshed between 50 and 200 times per second. The display mode can be arbitrarily selected in the modern universal analysers.

The *timing display* is similar to the display mode of oscilloscopes, each input signal being represented by a series of square-waves on the screen. The parallel placing of the individual signals makes the relative timing measurements easier.

Another display mode is the *state display*, where the stored data are displayed in binary format or another different code. Figure 6.2 shows the binary display using 0 and 1 symbols. The state display is most convenient when tracing software faults is being carried out.

Another display mode of logic state analysers is the *map display*, where each input state is represented by a brightened-up dot on the screen (Figure 6.3). The X coordinate of the dot is determined by the lower-magnitude bit of the displayed word; the Y coordinate by the higher-magnitude bit. The intensity of the dots indicates the frequency of data change, while the lines connecting the dots indicate its direction. The operation of complex digital apparatus can be tested easily and simply using such map displays.

Figure 6.2 State display on the logic state analyser screen

Figure 6.3 Map display on the logic state analyser screen

One of the most useful display modes of logic analysers is the mnemonic disassembly. In this mode the measured data appear on the screen in the same form as they are written in the program.

6.2.1 The Tektronix Type DF2 display controller

The Tektronix Type DF2 display controller is an auxiliary unit of their Type 7D01 16-channel logic state analyser. The units plug into a standard 7000 Series oscilloscope mainframe and the task of the DF2 unit is to generate an IEC bus oriented display on the cathode-ray tube screen (Figure 6.4).

The DF2 unit can generate seven different kinds of display on the screen: timing, map, binary, octal, hexadecimal, ASCII and IEC 625 GPIB. The first six of these are universal, suitable for testing any digital circuit. The IEC 625 GPIB display, however, is specifically created for testing IEC systems.

In this operational mode a special adaptor equipped with an IEC interface must be connected to the input of the logic analysers instead of the usual measuring probe system. This adaptor switches the DIO1...DIO8 data lines of the IEC bus and the ATN, SRQ, REN and EOI control lines to the appropriate analyser inputs. There are also four free inputs, the measuring heads of which can be connected to any signal lines or circuit points by the user. These inputs can be used for testing the data transfer controlling DAV, NRF and NDAC signal lines, for example.

Figure 6.5 shows an IEC 625 GPIB display. From the 254 × 16-bit memory of the analyser the data from 17 sampling cycles are displayed simultaneously. This 17-row 'window' can be shifted by a cursor control located on the front panel of the analyser, and the user can thus scan through the entire contents of the memory. The numeric value in the

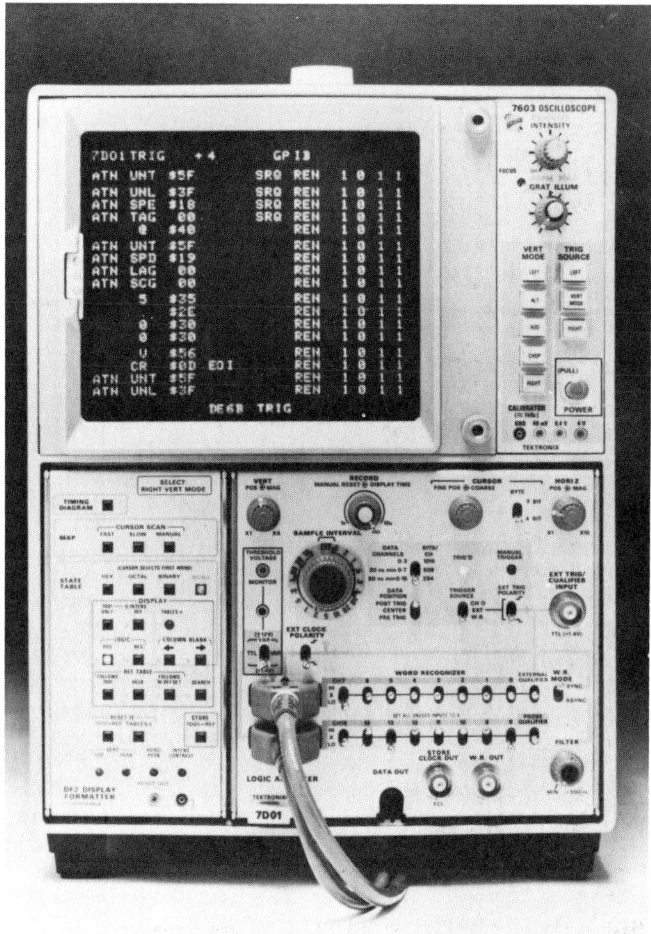

Figure 6.4 Tektronix Type DFZ/7D01 logic state analyzer

first line of the display (133 in our example), indicates the serial number of the sampling that the display begins with.

The basic information displayed on the screen is the state of the IEC bus data lines, using various code formats. The horizontal rows in the Figure represent the individual sampling cycles. The first four of these contain device-dependent messages. The individual messages can be seen next to each other in ASCII and hexadecimal code, e.g. 0;30 and CR;0D. The interface messages are displayed using a mnemonic format of the abbreviated message name (ATN, UNL, REN, etc) and this makes the interpretation of the data series seen on the screen much easier. The talker and listener addresses given by the T and LAG messages are displayed in decimal format. In these rows the introductory character identifying the hexadecimal format is not used. The right-hand side of the screen shows the binary format values of the four signals that can be selected by the user.

The Type DF2 display controller can be used for automatic fault indication. For example, if an erroneous (non-identifiable) interface message appears on the bus in the true state of the ATN message, then this erroneous line is displayed with increased intensity.

The display mode outlines above can be used only in synchronous operational mode. The external clock signal can be the negative going, trigger edge of the DAV signal, for example. This ensures that the analyser will sense the data transmitted to the bus as the IEC devices do.

6.2.2 The Hewlett-Packard Type 10051A IEC adaptor

The Type 10051A adaptor is a special auxiliary unit of the Hewlett-Packard Type 1602A logic state

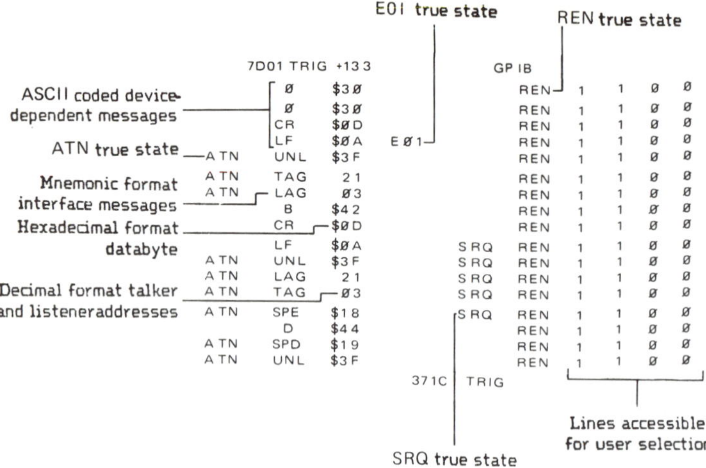

Figure 6.5 Data format obtained in the GPIB operational mode with the Tektronix DF2 display controller

analyser. The 10051A/1602A system is fundamentally different from the Tektronix Type DF2/7D01 system in that the special IEC adaptor is not located in the display section, but in the logic state analyser input, i.e in the data collecting section.

This is not the only difference between the two measuring systems. This analyser does not have a screen, the display appearing on a 16-character alphanumeric LED line. This display shows the contents of the 64 × 16-bit capacity memory line-by-line, in binary, octal or hexadecimal format according to choice. If the analyser is used for testing an IEC system, a plastics sleeve can be placed under the display to identify the IFC, SRQ, ATN, REN and EOI interface lines.

The task of the Type 10051A adaptor consists of transferring the IEC bus signal lines to the appropriate analyser inputs and the selection of the data. The selection is carried out with the qualifier signal. If the ATN signal is used as the qualifier,

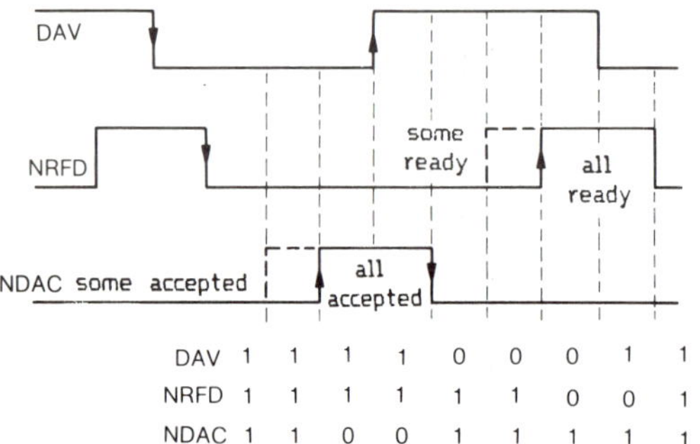

Figure 6.6 Simplified timing diagram of the handshake process

then according to the setting of the input selector, the analyser senses either the interface messages, or the device-dependent messages, or both.

The selection of the signal controlling the analyser's sampling requires special attention. There are four possibilities for IEC system testing. The data collection can be controlled with either the positive or the negative going edges of the DAV signal, or with the negative going edge of the NRFD signal, or with the positive going edge of the NDAC signal. The timing relationships of the IEC bus data transfer cycle are shown in Figure 6.6. If the analyser is clocked by the negative going edges of the DAV or NRFD signals, then the data appear in the memory prior to the actual data transfer. If the sampling is controlled by the positive going edges of the DAV or NDAC signals, then the data that have been accepted by the listener devices appear in the memory.

One of the main advantages of the Type 10051A/1602A system is its suitability for automatic testing of the handshake data transfer cycle. In Figure 6.6 the sequence of the DAV, NRFD and NDAC signals are shown for a correct data transfer cycle. This fixed sequence permits automatic testing of the data transfer. In Figure 6.7 the continuous repeated cycle of the handshake data transfer is sketched out, indicating the signal levels of the data transfer controlling bus in each state and the changes between the individual states. The transfers denoted with broken line arrows in the Figure represent an acceptable transfer cycle. In interpreting the Figure, it must be remembered that the IEC system uses a negative logic state level assignment.

In the Type 10051A adaptor, shift registers and comparators automatically and continuously monitor the data transfer cycle. If there is a deviation

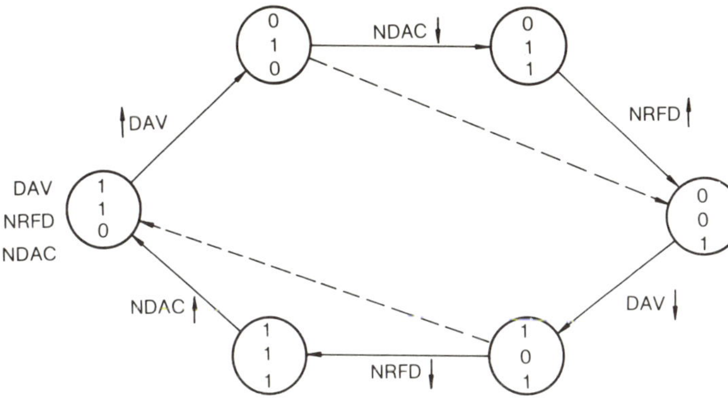

Figure 6.7 Normal sequence of the handshake process states

from the normal sequence, it is indicated by a 50-msecond-duration flash of an LED marked HANDSHAKE, located on the adaptor. At the same time the adapter sends a trigger signal to the analyser, thus starting the storage cycle at the first appearance of the error.

6.3 Active IEC system analysers

Table 6.1 shows the characteristics of some active IEC system analysers. These instruments differ from the logic state analysers described previously in that in the talker mode they are capable of transmitting any remote message; in the listener mode they can slow down the operation of the system to such an extent that the data traffic can be tested conveniently.

Of the instruments listed in the Table, the Hewlett-Packard Type 59401A is the most versatile; its photograph is shown in Figure 6.8 and its block diagram in Figure 6.9. The operation of the instrument can be understood from the block diagram in its two operational modes.

6.3.1 Listener mode

In the listener mode the Type 59401A can be regarded as a device in the 'listen only' state, accepting the entire data traffic of the bus without being addressed. The asynchronous data transfer allows the deceleration of the data traffic. The Type 59401A can operate at three different listening speeds. In the 'fast' setting the device does not influence the operational speed of the system; in the

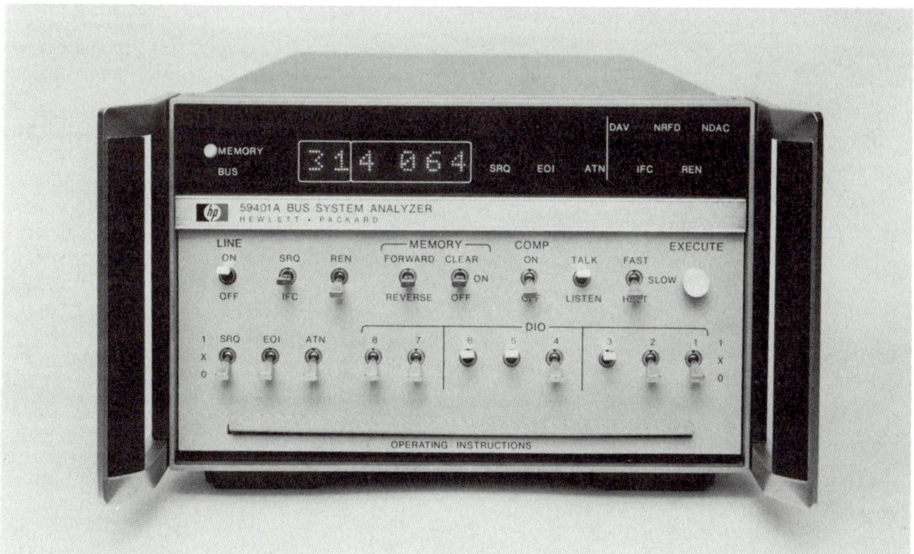

Figure 6.8 The Hewlett-Packard Type 59401A IEC system analyzer

Table 6.1 Active IEC system analysers

Type *Manufacturer*	*59401A* *Hewlett-Packard*	*4810* *ICS Electronics*	*ZT488* *ZIATECH*	*488* *Interface Technology*	*LA-1910* *Trio-Kenwood*	*GPIB-400* *National Instruments*	*488* *Racal–Dana*
Memory capacity, bits	33 × 16	100 × 16	–	54 × 16	1024 × 16	–	40 × 11
Display mode	ASCII octal	hexadecimal	binary	binary octal hexadecimal	binary hexadecimal	binary	hexadecimal
Combinative	yes	yes	no	yes	yes	no	yes
Operational mode (according to speed)	● stepped ● 2 byte/second ● external clock signal	● stepped ● 2 byte/second ● external clock signal	stepped	● stepped ● 250 kbyte/second	● stepped ● external clock signal	● stepped ● 20 kbyte/second	● stepped ● 100 kbyte/second

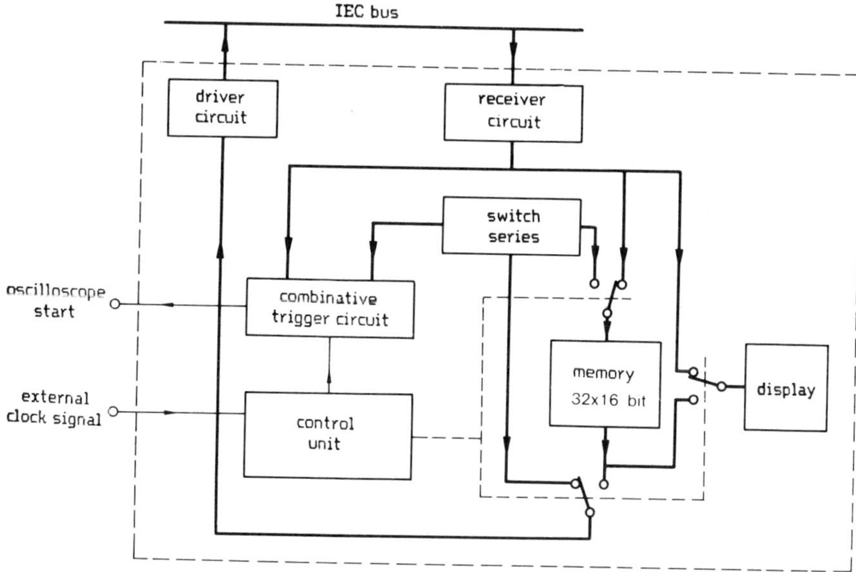

Figure 6.9 Structure of the Type HP59401A IEC system analyser

'slow' setting the speed is reduced to two characters per second; while in the 'halt' setting the data traffic of the system is manually controlled by the user operating the 'execute' button. The three speeds can be used alternatively and the user can switch over from the 'slow' setting to the 'halt' setting when the software error location is being approached during the execution of the program. In the 'compare' setting the instrument is automatically switched over to 'halt' setting when a pre-selected character appears on the data bus. The selection of the character is carried out with the series of switches located on the front panel. The combinative trigger unit, sensing the selected character, generates a trigger signal, which can be used for triggering the oscilloscope or the counter.

In the listener mode the data arriving on the bus are automatically transferred to the 32 × 16-bit RAM of the device. The storage takes place in the stepped sequence and the signals of the 32 cycles preceding the given characters will be found in the memory when the device is halted. This stored information can be displayed on the device in steps.

The display is simple and clearly laid out. The momentary states of the five controlling signal lines are indicated by the LED row, while the information present on the data bus is displayed in both octal and ASCII format, adjacent to each other. If the contents of a memory are displayed, then the memory address is also displayed with the data.

6.3.2 Talker mode

In the talker mode the device can send data and messages to the IEC bus at various speeds. The source of the data is either the switches located on the front panel, or the memory. If the data input is carried out from the front-panel switches, then only the 'halt' operational mode is possible. However, if the data is sent to the bus from the memory, the operation can be executed at the internal two characters per second speed, or at any arbitrary speed, using an external clock signal.

The latter solution is very advantageous for testing the listener functions of the system's devices, provided that an appropriate control program is loaded into the memory of the Type 59401A, and it is transmitted to the bus at different speeds. By increasing the frequency of the clock signal originating from an external source (e.g. impulse generator), the faulty listener device can be identified.

It must always be remembered that active system analysers represent normal device loading on the system under test. Accordingly, they cannot be connected to a system that already contains 15 devices.

7 Interface and backplane bus standards

Introduction

In earlier chapters we have considered the various classes of system hierarchy and briefly reviewed a few of the standard interfaces in common use. In particular we have studied the IEC 625 recommendations, whereby individual, programmable instruments may each be interconnected to other similar instruments and to a controller, thus forming an integrated system based on the serial transmission and handling of data bytes.

There are, however, a number of approaches open to us when designing automatic measuring systems that are based on digital processing. In particular a considerable number of interfaces and backplane buses are now commercially available; the merits of these vary considerably, often being somewhat dependent upon the applications. These buses are often traceable to specific designs of microprocessor chips and frequently have a high degree of commonality in their features. In most cases there are only minor differences in the electrical signals, timing and functional requirements of these various buses, but these are sufficient to prevent any common application of the different component boards by a user, who can therefore easily become entrapped with a particular supplier. Such a course can have expensive consequences, depending upon the technical and financial resources of the supplier and his ability to provide proper levels of hardware and software support after purchase.

Furthermore, only a few of these buses have been accepted as international standards, although several more are under consideration for adoption; additionally several are the in-house standards of companies of international repute and therefore are effectively *de facto* standards.

Whilst it is not our intention to carry out an extensive treatise on this topic, a knowledge of bus specifications is necessary so as to be able to configure a system using many modules. We therefore consider briefly the most important features before reviewing the more commonly available interfaces and backplane buses as well as those that have already received or are being considered for acceptance by the international standards organizations.

7.1 Principles of interfaces

The act of connecting two modules together creates the necessity for communication paths and therefore establishes an interface. Within each module the components will communicate via the internal wiring in whatever form this exists, but for the modules to interact in a systemized manner some common intercommunication technique must be used.

These techniques fall into one of two broad classes: *serial* or *parallel interfaces* (*See* Section 7.3). In minicomputer applications the latter are commonly called 'backplane buses' when applied to self-contained computer systems.

7.1.1 Serial interfaces

Serial interfaces require only one or two signal lines and data are transmitted as a stream of bits over one line. This has an advantage over parallel interfaces in that the number, and therefore the cost, of drivers and receivers to buffer the signals and process them is significantly reduced as is the complexity of the interface connections. Converters from parallel to serial and back again are required in those applications where the data are presented or required in parallel form. Bi-directional lines for the transmission of both data and control signals are also available.

Serial interfaces connect commonly-used peripherals such as VDUs, printers and similar devices to the systems allowing the bit-by-bit transmission of data, addresses and control signals. They are also frequently used to transmit bus functions in networks. The relevant standards define features such as the data transmission rate, data format and electrical characteristics.

Interfaces often use an interlock handshake over the data flow control lines to confirm that data are only sent when the receiving device is ready and to

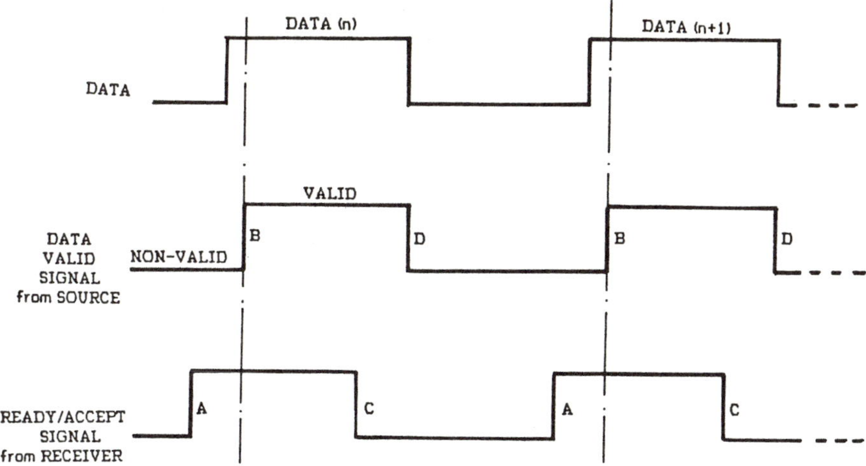

Figure 7.1 Handshake signals and timing sequence. Notes: A = Ready to accept DATA, C = DATA accepted, A, B, C, D = Full handshake sequence

confirm its correct receipt. Data handshaking is a sequence of interlocked signals ensuring that each device waits for an acknowledgement of its last transmission before proceeding to the next stage. This protocol is normally a software function as is parity generation and checking. The technique, which is illustrated in Figure 7.1 for a point-to-point application, such as an interface line in a radial configuration (*see* Figure 7.3), can also be applied to parallel buses.

In all system applications care must be taken to cater for the condition where no response to a transmission command is received. To ensure that the entire system does not become locked up (hung up), and to indicate the occurrence of an error, provisions, such as a timer mechanism, must be made.

The principal disadvantages of serial interface systems are a slow rate of data transfer, especially when compared with parallel transmission, the need physically to change wiring should the system require alterations, and its inherent inability to incorporate features such as DMA.

Serial interfaces may be either synchronous or asynchronous (including isochronous). In the former case the data bit rate is determined by clocks in the transmitter and receiver and the data is therefore transmitted at a fixed rate. The clocks must be kept in synchronism with each other by transmitting a clock synchronizing signal with the data (i.e. self-checking coding), or over a separate line.

Asynchronous systems are commonly used for low-speed terminals (<1200 bits/second). They operate only when data are to be transmitted and the data characters are preceded and followed by start and stop framing bits. Such systems tend to be less efficient in their use of bus space than synchronous transmission because of the extra start and stop bits and the minimum time necessary between characters even when using isochronous transmission.

7.1.2 Serial interface standards

The most common serial standards relevant to our interests are the ASCII and EBDIC standards and the CCITT V24 (EIA RS-232-C) synchronous standards. The ASCII code has already been mentioned in Chapter 2, and is shown in Table 2.19. ASCII and EBDIC asynchronous transmissions are commonly used by most VDUs and teletypewriters. Eight serial data bits, preceded by the start bit and concluded by the stop bit(s), form a character and are set over a cable. By the use of inexpensive parallel-to-serial and serial-to-parallel converters, i.e. universal asynchronous or synchronous receivers/transmitters (UART and USART), a simple twisted pair or a coaxial cable can be used for links up to one or two kilometres in length at data rates of a few kbits/second. For longer distances modems must be used.

RS-232-C (V-24) has been only briefly referred to in earlier chapters. However, although it is slowly being replaced by later standards it is discussed more fully here.

7.1.3 The RS-232-C (V-24) interface

The Electric Industry Association (EIA) Standard RS-232-C is the most commonly used interface for bit-serial data transmission and is the most frequently available port provided in small computer systems. It is virtually identical to the CCITT Rec. V.24. interface standard when used for these applications and both were originally established to define the electrical, physical and functional interface requirements of telecommunication authorities for connecting Data Terminal (DTE), Data Control Equipment (DCE) and modulators and demodulators (MODEMS).

It is used as an asynchronous serial interface where data is transmitted over a pair of wires using nominal voltages of +12 V and −12 V to represent 'mark' and 'space' respectively. Additional lines are used to carry data flow control handshaking and other control signals and a 25-pin connector is defined. This is a 'D' type connector and the most common pin allocations are indicated in Table 7.1.

Table 7.1 The most common RS-232-C pin allocations

Standard reference RS-232-C	V24	*Pin number*	*Function*
AA	101	1	Enclosure Ground
BA	103	2	Transmitted Data (TD)
BB	104	3	Received Data (RD)
CA	105	4	Request To Send (RTS)
CB	106	5	Clear to Send (CTS)
CC	107	6	Data Set Ready (DSR)
AB	102	7	Signal Ground
CF	109	8	Data Carrier Detect (DCD)
CD	108/2	20	Data Terminal Ready (DTR)
CE	125	22	Ring Indicator (RI)

Notes:
25 pin, D-Type Connection.
Signals on Pins 2 & 3 (plus 7) are those necessary for the minimum RS-232-C requirements.

In practice, for the simplest bi-directional links only the two lines, Transmitted Data and Received Data, are needed in addition to Signal Ground. Handshaking lines, Request to Send, Data Flow Control and Clear to Send are desirable but not always supported in all microcomputers.

Data transfer rates up to 20 kbits/second are possible and standard data rates are:

50, 75, 110, 150, 350, 600, 1200, 2400, 4800, 9600, 1920 bits/second

Data transmitted over RS-232-C interfaces may use any convenient code, but ASCII is commonly used for computer systems.

EIA RS-232 has been supplemented by EIA RS-449, RS-422 and RS-423. These later standards are designed to permit higher data rates (≤2 Mbits/second) and the use of longer lines. RS-449, which corresponds to a subset of CCITT Rec. V.24, defines the mechanical and functional characteristics for synchronous and asynchronous serial binary data communication systems. Electrical characteristics for balanced and unbalanced circuits are specified in RS-422 (CCITT Rec. V.11) and RS-423 (CCITT Rec. V.10) respectively, and a 37-pin connector is used.

Despite these more recent standards, RS-232-C has such extensive use that it seems likely that some considerable time will elapse before its eventual demise.

7.1.4 Current loop

In some cases where transmission is required over longer distances or in particularly noisy environments, it is also possible to use a current loop system. This consists basically of a total of two pairs of wires, each pair carrying serial data in Full Duplex Traffic mode (FDX). A loop possesses a transmitter at one end and a receiver at the other. The arrangement is reversed for the other loop, thus permitting the system to exchange information. In addition, a current generator is necessary in each loop and this may be incorporated in the transmitter, which controls the levels of the current. Additional receivers can be added into the system without complexity.

There is no clearly-defined and internationally-recognized standard available for current loop systems; however, the currents are normally 20 mA and the system is therefore often otherwise known as the '20 mA current loop'. Optocouplers are available from several manufacturers and are invariably used, allowing ground loops to be eliminated and common mode rejection to be greatly enhanced. Data rates of 20 kbits/second and transmission distances of 10 km are possible using these devices over suitable cables. Teleprinters normally operate with a 20 mA drive current and the 20 mA current loop sees frequent usage for this purpose.

7.1.5 Hewlett-Packard Interface Loop (HP-II)

In conclusion of our review of serial interfaces we should mention the Hewlett-Packard Interface Loop (HP-IL). This is a bit-serial interface particularly designed for low-cost systems, which may be comprised of battery supplied devices. It is designed to complement the IEC 625 (HP-IB) interface system that has been the main topic of this book.

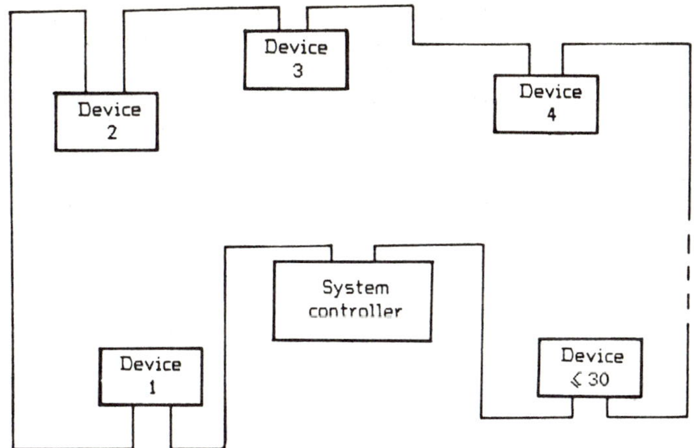

Figure 7.2 Hewlett-Packard interface loop

The HP-IL system makes use of a computer- or microprocessor-based device to act as a system controller and to manage the loop, controlling the devices connected to it. Up to 30 devices can be connected in the loop, as shown in Figure 7.2. Connections are effected by two-wire cables, running from the output port of one device to the input port of the next, the whole assembly comprising a closed loop as shown. The arrangement permits addresses, device capability identification, power on/off and error checking to be performed.

Of these, only the device capability identification perhaps needs clarifying. Each device possesses a unique capability number that identifies to the system controller the features that the device can undertake, e.g. 'print', etc. In executing a 'print' command, the controller polls each device to find the one that will respond with the appropriate identity. This procedure avoids the need for the user to know the address of each device and also simplifies the production and operation of system software and management, enabling devices of different speeds to operate in a unified manner.

The system is not the subject of an international standard. An increasing number of devices, varying from a hand-held computer to a digital multimeter, are manufactured to operate with the system. Interfaces are available to connect to other systems such as IEC 625 (HP-IB) and RS-232-C.

7.2 Data-link control protocol

Protocol is a set of rules to be followed in operating a communication link or system. It covers aspects such as the detection and correction of errors, the sequence of message transmissions, the framing of characters, the control of the transmission line, etc. A number of protocols are in common usage including the ISO High-Level Data Link (HDLC) and the more commonly met IBM's Synchronous Data-Link Control (SDLC).

7.2.1 Synchronous data-link control

Synchronous serial transmission has a number of sophisticated protocol variations under the general heading of Synchronous Data-Link Control (SDLC) when used in distributed systems requiring faster data links. Data are sent without start or stop bits being added to each character and means are required to avoid the risk of the link receiver losing its timing when data are sent intermittently in this way. This is achieved by adding synchronizing characters every hundred bits; as these characters are decoded by the receiver, they can be used to keep its clock in synchronism with that of the transmitter. In SDLC practice, a number of transmitters and receivers may be connected to a link and data transmitted in large blocks of characters, called 'frames', each of which has a number of fields of one or more bytes of data. The start character indicates to all the receivers that a frame transmission has begun and an address follows. The appropriate receiver accepts the frame, checks it and returns a frame to confirm its accuracy or to highlight an error that has been detected.

A standard defines the protocol and covers such aspects as check characters, types of frame, system start/stop and error characters. Integrated circuits providing all the necessary protocol and other features are available from several manufacturers.

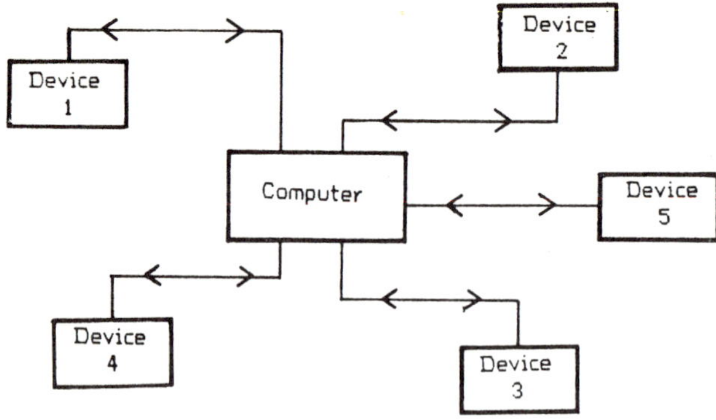

Figure 7.3 Radial or star connection

7.3 Parallel interfaces and buses

In parallel interfaces the bits comprising data words are transmitted simultaneously over the interface on parallel lines. One line is used for each bit and an interface may be one or more words in width. Because the transfer is a total transfer of the word in a single cycle, the parallel interface operates in a fast manner, particularly in comparison with serial interfaces. However, in contrast to the serial interface, transceivers are required for each line.

Computers may use a radial or star form of interconnection, as illustrated in Figure 7.3, where each device connection is achieved by a dedicated, parallel interface. Such an approach is expensive for the reasons mentioned above, and is now used only rarely. A more common method is the 'daisy chain', or 'highway', illustrated in Figure 7.4. With this technique, only one drive and receiver are necessary in the processor.

However, in this section we wish to concentrate on the use of the parallel bus as commonly used in systems and this is illustrated in Figure 7.5. These buses are designed to provide a common means of communication between a wide range of system modules, such as single board computers, memory, digital and analogue I/O and controllers for peripheral devices. The bus structure must therefore accommodate all the necessary signals to allow the various system components to interact with each other.

7.3.1 Parallel bus

The parallel bus transfers all the bits in a data word simultaneously across individual lines. Separate groups of these lines are allocated for the data bus, address bus and control bus and a typical, simple microcomputer will comprise 8 data, 16 address and 5 (or more) control lines. In some cases, and in particular, in high-performance systems, line capacity will be effectively increased by the use of multiplexing. Power lines will also be required. All this is achieved by the use of a mother board technique, which forms the bus and comprises the bus lines, together with a means of mechanically accommodating and supporting the individual modules or device boards. This system is often termed a 'backplane bus', and a basic example is shown in Figure 7.6.

The *data lines* are used for the transfer of all data in and out of the processor(s).

The *address lines* identify the area of memory or I/O the transfer for which it is intended.

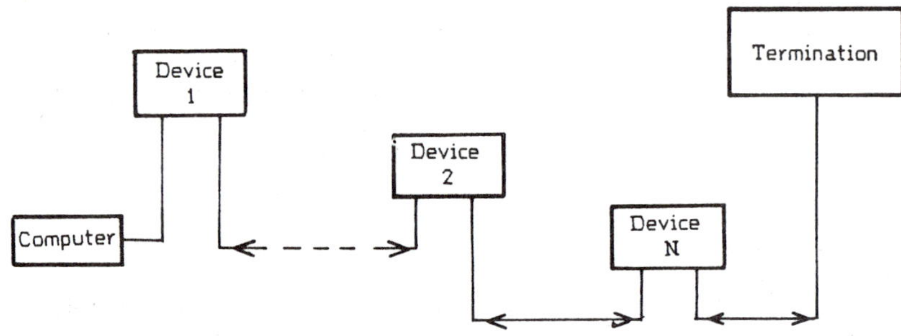

Figure 7.4 Daisy-chain or highway connection

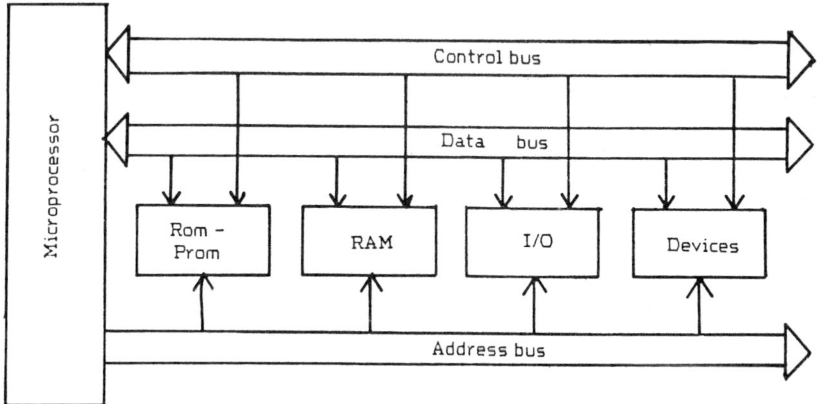

Figure 7.5 Typical parallel-bus system

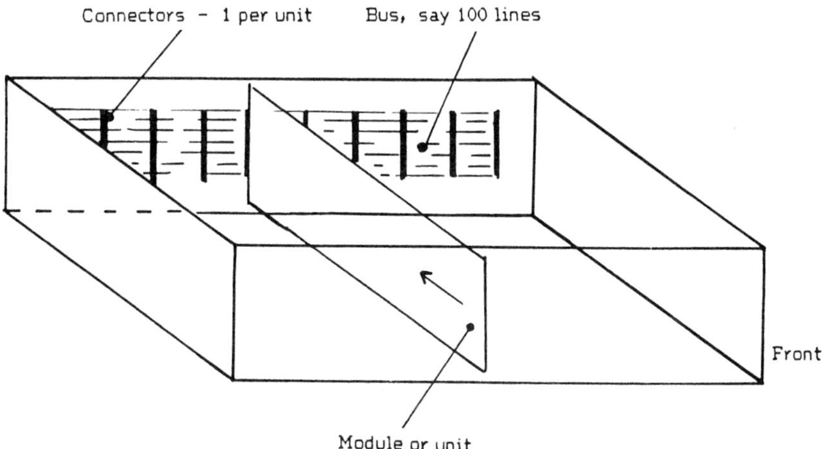

Figure 7.6 Schematic physical assembly of a backplane bus

The *control lines* typically carry read or write instructions, address ready, wait or respond signals, an interrupt signal or a DMA request.

Depending upon the particular system, the *power lines* may be common, with regulation provided at each device, or individually-regulated to each location.

7.3.2 Multiplexing

The number of bus lines required for data and address can be reduced by the use of multiplexing. Data and address information share the same bus lines in some prearranged sequence. For instance, address bus lines A_8 to A_{15} can carry the most significant 8 bits of the memory or I/O address, whilst the multiplexed data/address bus lines A_0 to A_7 carry the least significant 8 bits during the first clock cycle. Following this during the subsequent two or three clock cycles, lines A_0 to A_7 become the data bus.

A multiplexed system thus requires two distinct transmissions together with synchronization facilities. Multiplexing has some cost and reliability benefits by virtue of the fewer components required but the multiple transmissions slow down the system response.

7.3.3 Word length

An important feature of backplane buses is the size of word that they accommodate. The range of small computers presently available commonly use either 8- or 16-bit words and the usefulness of a backplane bus is very dependent upon its ability to support either or both of these. 32-bit processors are now becoming available and these can be supported by only a small number of buses currently available.

7.3.4 Interrupt priorities, multiprocessors and arbitration

The priority aspect of backplane buses is important because it determines which of the various devices on the bus can interrupt to gain control of the bus and at what time. Simple priority schemes depend upon the location of the cards in the bus relative to the microprocessor, those nearest receiving the highest priority.

More elaborate schemes are necessary for the more sophisticated buses which permit several 8- to 32-bit processors to share the bus. Such schemes generally use one microprocessor to act as a bus supervisor, monitoring requests from the various processors. These requests are made on separate bus-request lines and the grants made on corresponding separate bus-grant lines.

A priority system such as this requires a mechanism to arbitrate between simultaneous claims for control of the bus. Multiprocessor buses undertake this by queueing the requests – generally on the basis of sequential servicing; however, devices can share cycles on the bus or a particular device can take priority under stated conditions. Overall no one device can keep permanent control of the bus and if necessary, requesting devices can make use of a 'protest' line to initiate a software control response.

7.3.5 Errors and their detection

Information on the bus may become degraded by spurious signals from external circuits, crosstalk on the bus lines, mismatch in the bus, etc, leading to deterioration in an apparently satisfactory system.

Error detection and correction is necessary to prevent the complete disruption of the system and may be implemented by careful hardware design or by software. This may take the form of well-defined data test patterns, use of lateral or longitudinal parity checking on each data word or block, or by cyclic redundancy checks. The latter consists of carrying out a repetitive calculation on the data sequence being transmitted at each end (transmitter and receiver) of the transmission path and comparing these. It is a particularly powerful method, permitting a very low undetected error rate.

7.3.6 Bus length and termination

The physical length of the backplane bus is related to its performance and operating speed. In general, most reasonable performance direct buses are considerably less than 1 m in length and the propagation time of the signal along the bus lines is highly relevant. In all cases, the bus may be regarded as a transmission line and proper terminations are necessary to avoid malfunctions due to mismatch, reflections, ringing, etc.

7.3.7 Bandwidth

Most backplane buses have an effective bandwidth, depending on the physical properties of the bus, of between 5 and 10 MHz. In practice, the transfer rate of data is more important, since a word may consist of 8 bits or multiples of 8 bits for each operating cycle, and this is expressed as Mwords/second.

7.4 Backplane bus standards

In the reviews that follow, a number of the more common and important backplane bus standards are described and their principal features summarized. The treatment is not exhaustive and the reader is referred to the specification documents for the full and usually comprehensive information. A number of abridged specifications contained in the following sub-sections are based on information from the journals *Microprocessors and Microsystems* and *Electronics Industry*, and the permission of the publishers to use this is gratefully acknowledged. The omission of some buses is not intended to imply any criticism, but merely indicates the magnitude of the problem facing the prospective system designer.

The backplane buses surveyed are:

S100	(IEEE 696)
STD	(IEEE P961)
STE	(IEEE P1000)
G64	–
Eurobus	(ISO-DP 6951)
Multibus I	(IEEE 796)
Multibus II	
VME	(IEEE P1014)
Futurebus	(IEEE P896)

The information given is necessarily brief – full details are given in the specifications available from the appropriate standards organizations or design authorities, whose addresses are given in the Bibliography section.

7.4.1 The S100 backplane bus (IEEE 696)

The S100 bus, otherwise known as the IEEE 696 bus, is an example of how an informal, loosely-specified bus proposal can become widely accepted. Despite much initial incompatibility from a wide range of suppliers, the specification was refined and eventually, when supported by a respected standards organization, has become a usable tool.

The S100 bus comprises 100 bus lines which include 15 bi-direction data, 16 or 24 address, 8 state, 11 bus control, 6 DMA control and 8 vectored interrupt lines. It supports 8- and 16-bit processors, permitting up to 16 Mbytes of memory and 64 k I/O ports to be addressed. Its specification is such that almost any 8- or 16-bit processor from on 8080 onwards can be accepted and a data transfer rate of 6 Mwords/second is achieved.

The devices are mounted on standard cards (254 × 135 mm) and use 100 way printed circuit edge connectors. Up to 22 slots are available to accommodate from 4 to 22 device cards.

Every system must have a permanent bus master, which is usually the CPU together with memory and I/O slave boards. The system organization is such that up to 16 temporary masters may also be accommodated. Memory slaves and I/O cards have 24 and 16 address lines respectively.

8-bit transfers are made on two sets of uni-directional data buses, one for master-to-slave and the other for slave-to-master; 16-bit transfers on both buses are used together in a bi-directional mode and two extra bus lines are used for request and acknowledge. Both 8- and 16-bit slaves can therefore be mixed in a system subject to the controller possessing an enhanced byte serializer facility which serves four arbitration lines.

S100 abridged specifications

Type	Synchronous, non-multiplexed
Address field	16 Mbyte – 24 address lines
Data width	8 and 16 bits
Card size	254 × 152.4 mm (max)
Connector	Edge
Connection capacity	100 ways
Power supplies	±8 V, ±16 V (local regulation)
Interrupt levels	8 vectored
Bus handover	8 DMA
Arbitration means	4 lines
Bandwidth (transfers)	6 MHz
Maximum board capacity	22

7.4.2 The STD backplane bus (IEEE P961)

The STD bus emerged around 1978 and has been improved over the years. It is an 8-bit, synchronous, non-multiplexed bus and is specified so as to allow it to support any 8-bit processor family and simple multiplexer systems. It accommodates a standard printed circuit card (165 × 114 mm) with a keyed integral printed circuit edge connector.

The bus is comprised of a 56-line bused mother-board that permits any card to operate in any slot. Memory up to 64 k is supported on the bus and this can be expanded to 128 k by the use of the expansion line.

A European version of the STD bus (E-STD) has been proposed in order to foster more acceptability of the standard in Europe. This includes the provision of more address bits, use of a standard Eurocard (160 × 150 mm) and a 64-pin connector complying to DIN 41612.

STD systems are configured to accommodate a diverse range of devices. Peripheral and I/O devices are connected using their own unique front edge connectors and cable to the external device. No limit is placed on the number of slots on the bus but the backplane is sensitive to length and loading and each card represents one load per bus signal.

The extensive potential flexibility of the system requires caution in respect of compatibility. For instance, cards using peripheral chips usually depend upon specific timing signals for their satisfactory operation and this sometimes prevents their use with other processor families. STD practice is to label cards that are processor-dependent with a reference to the particular CPU, e.g. STD 6800, etc, and to specify any relevant waveform and timing requirements.

Interrupt servicing is available and is directly implemented by signal lines in systems with only a single interrupt device. In systems with a number of interrupt devices both serial and parallel schemes are defined. In the case of the European STD bus, request and acknowledge lines for either interrupts or alternatively, bus arbitration are provided. This permits up to two controllers to be accommodated if a daisy-chain technique is used for interrupts.

STD abridged specifications

Type	Synchronous, non-multiplexed
Address field	64 kbytes
Data width	8 bits
Card size	165 × 114 mm
Connector	Edge
Connection capacity	56 ways
Power supplies	±5 V, ±12 V
Interrupt levels	2
Bus handover	1 DMA
Arbitration means	None
Error signals	None
Special cycles	Interrupt acknowledge
Bandwidth (transfers)	2 MHz

Eurocard E-STD abridged specifications

Type	Synchronous, non-multiplexed
Address field	64 Kbytes
Data width	8 bits
Card size	160 × 100 mm
Connector	2-part DIN 41612
Connection capacity	64- or 96-pin
Power supplies	±5 V, ±12 V
Interrupt levels	3
Bus handover	Multiple master
Arbitration means	1 bus acknowledge, daisy-chain
Error signals	None
Special cycles	Interrupt acknowledge
Bandwidth (transfers)	2-3 MHz

7.4.3 The STE backplane bus (IEEE P1000)

The STE bus is a refined version of the STD bus. It makes use of Eurocards and is different from the STD bus in several fundamental ways.

The bus is intended for implementing high-performance 8-bit microcomputer systems either in a stand-alone mode or in a multi-bus architecture. It may also serve as a high-speed I/O channel. In contrast to the STD bus it functions asynchronously and is based upon the master/slave concept where any master, having gained control of the bus, can address slaves via an asynchronous interlocked handshake. It is therefore possible to construct systems that incorporate devices of widely different speeds. Multiple masters may also be implemented within a single system.

Two independent system address spaces are supported, allowing memory of 1 Mbyte physical address space and 4 kbytes of I/O. Two daisy-chains are defined; one for bus arbitration and the other for priority assignment to interrupting slaves. The system integrity has been improved with respect to the STD bus by the provision of a bus error line for monitoring data transfers and arbitration.

The bus makes use of single Eurocards with 64-pin connectors to DIN 41612 and up to 21 cards can be accommodated in the bus.

STE abridged specifications

Type	Asynchronous, non-multiplexed
Address field	1 Mbyte (extended address memory)
Data width	8 bits
Card size	Eurocard: 160 × 100 mm (option 160 × 233 mm)
Connector	DIN 41612
Connection capacity	64-pin
Power supplies	±5 V, ±12 V
Interrupt levels	4
Bus handover	Multiple master
Arbitration means	1 bus acknowledge, daisy-chain
Error signals	Bus error
Special cycles	Read/Modify/Write Interrupt acknowledge Block transfer
Bandwidth (transfers)	2-3 MHz
Maximum board capacity	21

7.4.4 The G64 backplane bus

The G64 bus was devised by GESPAC SA of Geneva in 1979, and a wide range of modules is available from a number of suppliers.

It is a bus capable of supporting 8- or 16-bit microprocessors and the bus design is unrelated to any special CPU family. It is able to provide either synchronous or asynchronous transfers and comprises a non-multiplexed 64-line bus. Standard single Eurocards with a DIN 41612 type connector are used and between four and 20 slots are available.

8- and 16-bit memory modules are available each being compatible with corresponding 8- and 16-bit processor modules, and memory addressing capability of 256 kbytes is possible. Up to 1 K × 8 or 1 K × 16 precoded field I/O addresses are also available and I/O modules can be used with either size of processor.

The bus is available as a two-layer device without proper terminations and capable of operating up to 1 MHz transfer rates, or as a four-layer bus terminated with 80 Ω, when more than 1 MHz is possible.

Three interrupt lines are available together with one interrupt acknowledge line and one daisy-chain, thus allowing non-vectored or prioritized vectored interrupts to be serviced.

It operates as a monoprocessor system with the ability to service multiple DMA sources. These are serviced by Request Grant Acknowledge, using the interrupt daisy-chain to establish priorities in multiple DMA configurations.

It usually operates in a synchronous mode for 8-bit processors and in an asynchronous mode for 16 bits. Either mode is possible with a mixed system.

A 32-bit version of G64 is under development (G128) and although making use of double Eurocards, it is otherwise planned to be compatible with existing G64 boards.

G64 abridged specifications

Type	Non-multiplexed: 8-bit synchronous 16-bit: asynchronous
Address field	16 (>256 kbytes)
Data width	8 and 16 bits
Card size	Eurocard: 160 × 100 mm
Connector	DIN 41612
Connection capacity	64-pin
Power supplies	±5 V, ±12 V
Interrupt levels	3
Bus handover	Mono-processor with multiple DMA
Arbitration means	1 bus acknowledge, daisy-chain
Error signals	Parity, Bus error
Special cycles	Interrupt acknowledge Read/Modify/Write
Bandwidth (transfers)	8-bit: ≤1 MHz 16-bit: >1 MHz
Maximum board capacity	20

7.4.5 The Eurobus backplane bus (ISO-DP 6951, BS 6475 : 1984)

Eurobus is a general purpose backplane bus, originally designed jointly by the UK Ministry of Defence and the Ferranti company. A chip set has also been designed to support the standard and is commercially available.

The bus was originally designed for a 16-bit data width and was intended to join together processors, memory and I/O devices in a variety of single and multiprocessor systems. The specification has now been extended to allow data widths of 8, 16, 24 and 32 bits to be supported; corresponding address widths of 10, 18, 26 and 34 bits are available with titles of Eurobus 10, etc.

The bus comprises a 64-pin connector covering the Eurobus 10 and 18 systems and two such connectors for the Eurobus 26 and 34 systems. Data transfer is asynchronous and fully handshaken and is multiplexed onto a single highway. 16 or 18 data/address lines are used together with nine control lines, one interrupt line and 11 power lines.

The system uses double Eurocards (233 × 160 mm) with DIN 41612, 64-pin connectors, up to 20 devices in a maximum bus length of 500 mm and the bus must be terminated at each end.

Eurobus is based on the concept of central bus arbitration. The bus control is removed from the processor(s) and allocated to a dedicated bus arbiter serving all requirements. Devices are allocated a master/slave relationship according to their need and their potential. The arbiter has complete control of the bus on the basis of the master being the device initiating a transfer request and a slave responding. Slave status need not be a permanent one.

Eurobus caters for two fundamental and separate operations – interrupt and data transfer. The interrupt operation is very basic and fast, thanks to the control arbiter which then performs any further system operations necessary. The protocol is designed to provide maximum speed, whilst still ensuring a handshake at every stage of the allocation and data transfer processes.

A particularly useful feature of Eurobus is its ability to allow two Eurobus systems to be linked together and operated as one. The system protocol caters for the inevitable conflicts where there are potential masters on both buses.

Eurobus does not support block transfers for reasons of reliability in its intended environment.

Eurobus abridged specifications

Type	Asynchronous, multiplexed
Address field	18 (10, 26, 34)
Data width	16 (8, 24, 32)
Card size	Eurocard: 233 × 160 mm
Connector	DIN 41612
Connection capacity	64-pin
Power supplies	+5 V
Interrupt levels	1 bus line to arbiter
Bus handover	Multiple masters
Arbitration means	2 (arbiter)
Error signals	Cycle abort, Reset, Power fail
Special cycles	Read/Write Read/Write/Hold Vector Read/Write/Vector/ Retain
Bandwidth (transfers)	3–4 Mbytes/second
Maximum board capacity	20

7.4.6 The Multibus (I) backplane bus (IEEE 796)

The Multibus backplane was originally devised by the Intel Corporation in 1974. After further development, it is now widely used for microprocessor industrial and systems applications and receives wide commercial support.

The bus makes use of an 86-pin connector and includes 16 bi-directional data lines, 20 standard (or 24 expanded) address lines, 18 control lines and eight lines of interrupts. The expanded addressing, when required, is accommodated by a 60-pin connector, which also carries other features such as power fail, memory initializing control signals and bus exchange lines. Power failure lines are optional.

The bus operates in an asynchronous, non-multiplexed manner and can support up to 1 Mbyte of direct addressing with 20 bits and 16 Mbytes using the expanded 24 bits. It can also address up to 64 K via I/O ports using 16-bit addressing. Both memory and I/O cycles can support 8- or 16-bit data transfers.

The boards used are the original Intel cards of 304.8 × 171.5 mm and make use of printed circuit edge connections.

Multibus is particularly used to interface the Intel 80/86 single-board computers and an extensive range of memory expansions, digital and analogue I/O boards and peripheral controllers.

The bus structure is based on the master/slave concept, where the master takes control, placing the slave address on the bus. The slave decodes the address before following its commands and a handshake between the two devices allows modules of widely different speeds to use the bus at data rates up to 5 Mwords/second.

The bus also allows multiple masters for multi-processing and provides control signals on bus exchange lines for connecting them either in serial daisy-chain priority or in parallel. In the latter case up to 16 masters may share the bus resources.

Multibus (I) abridged specifications

Type	Asynchronous, non-multiplexed
Address field	20 standard (1 Mbyte) 24 expanded (16 Mbytes)
Data width	8, 16
Card size	304.8 × 171.5 mm
Connector	Edge
Connection capacity	86 ways (+ 60 ways)
Power supplies	+5 V, ±12 V
Interrupt levels	8
Arbitration means	1 bus acknowledge, daisy-chain
Error signals	None
Special cycles	Lock
Bandwidth (transfers)	5 Mwords/second
Maximum board capacity	16

7.4.7 The Multibus II backplane bus

In late 1983 Intel announced a new 32-bit bus architecture named Multibus II. This is based on principles similar to those of the original Multibus (I) and extends the multiple-bus approach to now use five buses, which are processor-independent.

The extra buses are:

Parallel System Bus (iPSB)
Local Bus Extension (iLBX II)
Serial System Bus (iSSB)

The original I/O Expansion Bus (iSBX) and the multichannel DMA I/O Bus from Multibus I continue to be used.

The buses are interconnected by a common system interface that defines intermodule communication and data transfer protocol and thus allows system designers to choose the most appropriate combination of the five buses to satisfy a particular requirement.

In consequence, Multibus II uses a processor-independent open-system architecture, which is suitable for a wide range of system designs. It provides a 32-bit parallel bus with a capability of 40 Mbytes/second transfer rate and high-speed access to large-capacity remote (off-board) memory. In addition, the serial system bus permits serial access to be achieved at low cost.

The advantages of the multiple-bus approach over a general purpose bus are claimed to be:

(i) A specialized bus carries out its functions more efficiently.
(ii) Different functions can be carried out simultaneously on different buses.
(iii) The fast bandwidth of the general purpose bus is retained, thus providing an efficient bandwidth for communication and data transfer between processors.
(iv) Only the buses necessary for particular system requirements need be incorporated, thus reducing costs.

The bus properties are summarized as follows:

The *Parallel System Bus* is a general purpose, multiple-processor-independent, synchronous bus that supports data movement and interprocessor communication using 8-, 16- or 32-bit processors. Arbitration and execution are also supported. Space is provided for 32-bit memory addresses, 16-bit I/O addresses, 8-bit message addresses and 16-bit interconnection addresses. Sequential arbitration is provided for devices on a priority basis. Data transfers are clocked at 10 MHz and can be 8, 16, 24 or 32 bits wide. A burst transfer data rate permits a sustained bandwidth of 40 Mbytes/second. The bus uses a standard Eurocard with a 96-pin connector to DIN 41612 and up to 20 devices can be accommodated.

The *Local Bus Extension* functions as a fast processor execution bus which extends the local bus to remote memory. It possesses a 48 Mbytes/second bandwidth and a 12 MHz clock and provides local memory expansion, without arbitration, to 64 Mbytes, whilst supporting 8-, 16- and 32-bit processors. Up to six devices can be supported on this bus.

The *Serial System Bus* is a simplified version of the Parallel System Bus, being only 1 bit wide and running at 2 MHz. Accordingly, it is much cheaper and also allows up to 32 devices to be extended up to 10 m distant from the backplane.

The *Multichannel DMA I/O Bus* permits high-speed block data transfer between peripheral devices and single-board computers. It is an asynchronous bus, capable of supporting up to 16 devices of 8 or 16 bits with 16 Mbytes of memory to each device. It can support data transfers at 8 Mbytes/second at a distance of up to 15 m.

The *I/O Expansion Bus* allows the use of Intel small boards to provide local (on board) system expansion as and when necessary.

Some aspects of the Multibus II are still in draft form and under active discussion through the IEEE.

Multibus II abridged specifications

Type	Synchronous, non-multiplexed
Address field	32 (8, 16, 24)
Data width	32 (8, 16, 24)
Card size	Eurocard: 100 × 220 mm 233 × 220 mm
Connector	DIN 41612
Connection capacity	96-pin
Power supplies	+5 V, ±12 V
Arbitration means	6 lines, 2 level
Error signals	DC low, Protection bus error
Special cycles	Block transfer, Lock
Bandwidth (transfers)	Burst: 10 MHz for 32-bit transfer Single: 20 MHz for 32-bit transfer
Maximum board capacity	20

7.4.8 The VME (IEEE P1014) and Versabus (IEEE P970) backplane buses

The VME bus is the Eurocard implementation of Versabus and is therefore smaller but still very suitable for industrial applications. The two systems are architecturally very similar, and accordingly our description is restricted to the VME bus which was introduced and supported by Motorola, Mostek, Signetic/Philips and Thompson/EFCIS. The proposal is now out for public comment. An important attraction of the VME bus is the fact that it was fully specified before any cards were made available.

Whilst the VME bus harmonizes with signals of the 68000 microprocessor, other processors can be accepted. The specification allows for data transfers of words 8, 16, or 32 bits wide and the use of 16-, 24- or 32-bit byte addresses.

The VME bus is an asynchronous, non-multiplexed bus that can be visualized as four separate buses, namely

the data transfer bus, DTB,
the priority interrupt bus,
the DTB arbitration bus, and
the utility bus.

The specification also defines the module card size, connector type and pin disposition, power supplies, etc.

The *Data Transfer Bus* is a 16-bit data path capable of expansion to 32 bits by the use of a second, optional connector. A 24-bit address path, also expandable to 32 bits by the second connector, three control lines and six address-modifiers together with other control lines are also specified.

The widths of data words are identified by an address modifier code in the bus control signal. Accordingly, a variety of processors with differing data widths or address widths can be mixed in a single VME system, the bus control allowing efficient co-operation through selected common memory areas.

The *Priority Interrupt Bus* has seven interrupt lines and one daisy-chain interrupt acknowledge line. The current master will acknowledge interrupt priorities via the low-order address lines and vectors associated with interrupts are passed via the low-order data lines.

The *Data Transfer Arbitration Bus* possesses four bus request lines and four grant daisy-chain lines. Any master can hold one of four priority levels during normal bus operations; each of these levels has a request and grant line dedicated to it.

The *Utility Bus* is dedicated to timing signals, initialize and bus diagnostics, together with a 16 MHz system clock.

The *Interintelligence Bus*. In order to avoid the customary problem of slowing the system down by use of the Data Transfer Bus merely to allow communication between the processors (e.g. global memory) an interintelligence bus has been specified. This consists of a two-line synchronous serial bus capable of operating at 4 MHz between intelligent devices, thus permitting intercommunication to bypass the Data Transfer Bus. Protocol has provisionally been specified but has yet to be approved.

VME abridged specifications

Type	Asynchronous, non-multiplexed
Address field	16 Mbytes (4 Gbytes with extended address)

Data width	16 (32 bits expanded)
Card size	Eurocard: 160 × 100 mm 160 × 233 mm
Connector	DIN 41612
Connection capacity	96-pin (+96-pin expanded)
Power supplies	+5 V, ±12 V
Interrupt levels	8 assignable
Bus handover	Multiple masters
Arbitration means	Centralized 4-bus acknowledge, daisy-chain
Error signals	AC fail, System fail
Special cycles	Read/Modify/Write, Block transfer, Interrupt acknowledge, Address privilege levels
Bandwidth (transfers)	10 MHz
Maximum board capacity	20

7.4.9 The Futurebus backplane bus (IEEE P896)

Futurebus has been devised by the IEEE microprocessor standards committee after their work on S100 and Multibus I. It is intended to be a forward-looking, long-life, manufacturer-independent standard. The draft of Futurebus is out for public comment and some of the details quoted here may therefore be amended later.

Futurebus comprises a high-performance, 32-bit highway operating with an asynchronous, multiplexed protocol; 8- and 16-bit data transmission can be supported and an independent serial highway is also included. It is designed to accommodate multiple processors and makes use of a distributed arbitration control. Daisy-chain arbitration and interrupt acknowledge systems are not possible, since the system does not rely on any particular cards being available in the system configuration.

The bus may be visualized as four buses, thus:

- *Parallel Data Bus*, providing 32 multiplexed address and data lines together with timing and state lines.
- *Parallel Arbitration Bus*, providing control and priority lines which allow a single master to take control of the Parallel Data Bus simultaneously with data transfer.
- *Serial Bus*, consisting of a serial line and a check line to permit separate transfers in addition to those of the Parallel Data Bus. The Serial Bus is an optional feature and not essential for Futurebus operation.
- *Power Bus*, which distributes the 5 V supplies and provides ground returns.

The family of Eurocards is recommended with a preferred choice of the maximum size card (280 × 367 mm) together with a single 96-pin DIN 41612 connector. Up to 32 cards can be accommodated on the bus.

Any suitable task scheduling software should be satisfactory for use with Futurebus and permit the operation of multiple independent software systems. Synchronization of parallel processors is to be provided and under these conditions a burst data transfer rate of 10 Mwords/second (32 bits) is expected to be possible. For such transfer rates the bus must be properly terminated to avoid mismatches, etc, and to use the protocol efficiently. This is achieved by ensuring that addresses are already known to the slave, thus avoiding address transmissions.

The philosophy of the bus is that of a communication bus betwen processors. Fetch instructions, etc, are minimized and complete messages are passed. In consequence, the bus will support virtual memory systems for disks and cache memories.

Use of asynchronous protocol allows the speed of the current bus master to be achieved rather than slowing the system down to that of the slowest possible master.

A distributed arbitration system offers sufficient flexibility in the allocation of priorities for at least 32 prospective bus masters.

Futurebus abridged specifications

Type	Asynchronous, multiplexed
Address field	32 bits (4 GBytes)
Data width	8, 16, 32
Card size	Eurocard – preferred 280 × 367 mm
Connector	DIN 41612
Connection capacity	96-pin
Power supplies	+5 V
Interrupt levels	Virtual (by serial bus)
Arbitration means	32
Error signals	Error reports
Special cycles	Read/Modify/Write Block transfer Split cycle, Event cycle
Bandwidth (transfers)	10 MHz – 32 bits
Maximum board capacity	32

Glossary

AC Alternating Current.
access time The time required to complete a data transfer from a memory address, from the time when the request was made.
accumulator A general purpose register used for data storage and arithmetic operations.
ADC Analogue-to-Digital Converter – a device for converting signals from analogue to digital form.
address A pattern of characters identifying a unique memory location, peripheral, etc.
address bus A common electrical path carrying address data – *see* bus.
ALGOL A structured high-level language, particularly suitable for algorithmic applications.
algorithm A step-by-step procedural description to perform a defined function.
alphanumeric A set of characters comprised of letters, numbers and punctuation, etc.
analogue (signal) The representation of a variable parameter by a time-continuous voltage.
AND gate A circuit with two or more binary inputs and one output – giving a 1 output only when all inputs are 1s.
arbitration A mechanism to determine and allocate priorities to simultaneous interrupt calls.
ASCII The American Standard Code for Information Interchange – a 7-bit binary code representing alphanumeric characters.
assembler A program to translate programmes written in assembly language into machine code.
assembly language A programming language that allows symbolic labels to replace addresses and uses mnemonics to replace machine-code instructions.
asynchronous operation Operations (e.g. transmission of data) which do not depend on a commonly used clock or timing signal. The start of the next operation awaits the completion of the current one.
ATLAS Abbreviated Test Language for Avionic Systems. (IEEE 416).

backplane bus A system comprising a mechanical support mechanism and electrical bus connectors.
bandwidth The difference in frequency between the upper and lower limiting frequencies of a frequency band.
BASIC Beginner's All-purpose Symbolic Instruction Code – a conversational language.
baud A unit for the rate of data flow in serial transmission, e.g. 1 bit/sec = 1 baud.
BCD Binary Coded Decimal – a representation of 4-bit groups to each decimal digit in a number.
binary A method of representing numbers by 0s and 1s (e.g. bits). Successive bits in a binary number represent 1, 2, 4, 8, etc.
bit The basic binary unit in a processor; it possesses either of two values, 0 or 1.
block transfer The transfer of a group (block) of words.
branch driver The branch highway controller/driver in a multi-crate CAMAC system.
BSI interface British Standard defining the British Standard Specification 4421 interface.
BS 6475 Specification for a processor systems bus interface (Eurobus).
buffer A device to match different circuit impedances.
bus A parallel conductor system used to carry information to allow CPU, memory and I/O devices to communicate; commonly divided into address, control and data buses.
byte A collection of 8 contiguous bits.
byte serial, bit parallel The arrangement of data for transfer in the IEC 625 (IEEE 488) interface system.

CAMAC (IEEE 583 and IEC 482) An international modular interface system in which the mechanical, electrical and data-transfer interfaces are all standardized.
CCITT The Consultative Committee of International Telephone and Telegraph Union.
clear To restore a register or memory to a zero state.
clock (generator) A device producing timing pulses used for synchronizing computer operations.

CMOS Complementary Metal-Oxide Silicon – the most common unipolar semiconductor manufacturing technique.

CMR Common Mode Rejection – the rejection of fluctuations which cause both signal inputs to vary by the same amount, relative to signal ground.

code A system of representing data or directions by binary bits.

compiler A programme for translating a high-level programme such as FORTRAN to machine code.

complement The inversion of binary bits.

control bus A common electrical path carrying control data – *see* bus.

controller A device responsible for addressing devices in a (measuring) system and organizing the operation of the system.

CPU Central Processor Unit – the unit that controls a computer's interpretation and execution of instructions and arithmetic.

Crate A chassis (framework) capable of accommodating up to 24 CAMAC plug-in modules.

CRC Cyclic Redundancy Checks – a powerful method of checking for the presence of errors in data-transmission systems.

current loop A simple looped conductor system capable of producing good data transmission either over long distances or in electrically-noisy conditions.

cycle (time) A sequence of repeated instructions or events. (The time required to execute a set of events.)

DAC Digital-to-Analogue Converter – a device for converting data from digital to analogue form.

daisy chain A sequential interconnection of peripherals to a CPU in such a way that the location of the peripheral in the chain determines its interrupt priority.

data A collection of numeric, alphabetic or other characters representing information to be processed by a computer.

data bus A common electrical path carrying data; often multiplexed for bi-directional transmission – *see* bus.

data transfer The transmission of data from one part of a computer system to another.

dataway The common data interconnection bus used by the CAMAC system.

DC Direct Current.

devices The system units such as instruments, printers, etc, comprising an IEC 625 system.

digital instrument An instrument which measured (and displays) continuously-variable signals in discrete steps.

DIL (d.i.l.) Dual-in-line (package)–a packaging arrangement for integrated-circuit assemblies; plastic encapsulated with a row of pins at 2.5 mm centres down each side allowing the package to be directly fitted to a standard pcb.

DMA Direct Memory Access – a means of obtaining direct access to a computer's main storage without involving the CPU.

duplex The simultaneous transmission and reception of data.

EAPROM Electrically-Alterable Programmable Read-Only Memory.

EBDIC Extended Binary Coded Decimal Interchange Code – an IBM 8-bit pattern assignment.

ECL Emitter Coupled Logic.

EIA Electronic Industrial Association.

EPROM Erasable Programmable Read-Only Memory.

error The discrepancy between the measured and the true values of transmitted and received data.

ESONE European Standards Of Nuclear Electronics.

Eurocard The European standard printed circuit board.

execute To interprete and carry out a specific instruction.

ferrite storage A storage device using magnetic properties.

FET Field-Effect Transistor.

firmware Software instructions carried in a ROM.

flag A 1-bit register that gives indication that a particular condition has occurred.

flow chart A symbolic chart illustrating a sequence of operations to solve a problem.

FORTRAN FORmula TRANslator – a high-level language.

GDU (Graphic Display Unit) Commonly a visual display unit (VDU) capable of displaying graphical data in addition to alphanumeric.

GPIA General Purpose Interface Adaptor consisting of a purpose-designed IC for interconnection between the GPIB and a microprocessor unit.

GPIB General Purpose Instrumentation Bus – alternative name for the IEC 625 bus.

ground The common signal earth point.

half-duplex The transmission or reception of data.

handshake An interlocking communication between two parts of a system, to confirm the receipt of a data transfer.

hardware The mechanical, electrical and electronic parts of a computer system.
HDLC High-level Data Link Control – protocol for a sequential link level data transmission.
hexadecimal A numbering system based on the radix of 16 (0 to 9 and A to F) to represent all combinations of a 4-bit binary number.
high-level language A problem-orientated language similar to the user's language, e.g. BASIC, FORTRAN, ALGOL.
highway The main path for signals in a computer – *see* bus.
HLL *See* high-level language.
HP-IB Hewlett-Packard Instrumentation Bus.
HP-IL Hewlett-Packard Current (I) Loop.

IC Integrated Circuit – one which is implemented on a silicon chip and is equivalent to a large collection of discrete components.
IEC The International Electrotechnical Commission.
IEC bus The IEC 625 interface system – *see* GP-IB.
IEC 482 *See* CAMAC.
IEC 625 The IEC interface system for programmable measuring apparatus – *see* IEEE 488.
IEE The Institution of Electrical Engineers (UK).
IEEE The Institute of Electrical and Electronics Engineers (USA).
IEEE 488 The IEEE standard equivalent to the IEC 625 interface system
IEEE 583 *See* CAMAC.
instruction A programming statement specifying an operation and the location to which it is applicable.
instruction length May be a word length or multiple thereof; e.g. two-word instructions increase the directly addressable storage capacity.
interface Input and output associated control circuits between two devices, e.g. CPU and a peripheral.
interface, standard hardware A system providing standardized circuits or conditions to facilitate the interconnection of devices to a (computer) system.
interpreter A program which converts programs written in symbolic or high-level languages to machine code. It therefore facilitates user programming.
interrupt A mechanism whereby normal program operation is suspended to allow a special or sudden request from an external device to be serviced.
inverter A circuit with one input and one output and in which the output state is the inverse of the input.
I/O Input/Output – referring to devices or signals/data.
isochronous An activity occurring at regular intervals.
isolation voltage The maximum voltage that can be sustained safely between a device's I/O and ground.

language A collection of specific representations to allow information to be communicated between people and a computer as a written program of instructions.
LED Light-Emitting Diode – a semiconductor lamp.
listener A device requiring selective addressing and instructions before responding.
LSB Least Significant Bit.
LSI Large-Scale Integration (circuit).

machine code The lowest level of instructions, expressed in binary code, to which a processor can respond.
magnetic disk A disk with a magnetic surface on which data can be stored by selective magnetization of parts of the disk surface. The disk may be either rigid (hard) or floppy.
magnetic store Any device which can store data by selective magnetism.
magnetic tape A flexible plastics tape with a magnetic surface on which data can be stored by selective magnetism.
memory The part of the system that stores data and instructions, each item having its own unique address. Includes such items as disk and tape, etc.
message, uniline An IEC 625 message sent on one signal line only.
message, multiline The simultaneous transmission of two or more uniline messages.
mnemonic A means of representing operations, constants, program addresses, etc, by an alphanumeric name.
MODEM MOdulator/DEModulator – a device for coding and decoding digital signals on to, or from, a carrier frequency; commonly used to interconnect to telephone networks.
MOS Metal-Oxide Silicon.
MOSFET Metal-Oxide Semiconductor Field Effect Transistor.
monitor A program that monitors the operation of a computer system.
motherboard A backplane p.c.b. that accommodates the bus conductors and permits p.c.b. module cards to plug into p.c.b. sockets.
MSB Most Significant Bit.
MSI Medium-Scale Integration (circuit).

multiplexing The process of transmitting more than one signal (e.g. address or data) over a signal route or bus.

multiprocessor (systems) Systems which have a number of processors sharing the bus.

NAND gate A circuit block that acts as an AND gate followed by an inverter.

NOR gate A circuit block that acts as an OR gate followed by an inverter.

object program The final source language program after translation into binary machine code.

octal A number system using a radix of 8. Each digit position in the range of 0 to 7 represents a power of 8.

OR gate A circuit block with two or more inputs and one output, whose value is 1 when any input value is 1.

parallel data transfer The simultaneous transfer of all the bits of a word.

parity check A check to confirm the number of bits in a transmitted word whether it is odd or even (as previously designated).

p.c.b. Printed-circuit board – a flat insulating board which carries the items for a circuit comprised of discrete components. Interconnection of the components is achieved by electrically plating conductors on the surface as required.

peripheral A device connected to a computer system and controlled by the computer.

poll A procedure to identify the source of and allocate priority to multiple interrupt requests.

priority (interrupt) A means of providing certain interrupt requests with higher priorities.

program A series of steps or actions designed to achieve a desired result.

PROM Programmable Read-Only Memory.

protocol A set of values defining the presentation and interchange of information.

RAM Random-Access Memory – memory capable of being written to or read from in any order.

refresh To restore the necessary voltage drive of a dynamic storage unit.

register A temporary location with direct access by the processor.

ROM Read-Only Memory – memory to which data cannot be written by the system.

routine A sequence of program instructions for carrying out a specific task within a program.

RS-232 *See* V24.

SDLC Synchronous Data Link Control – protocol for synchronous data transmission.

semiconductor store Memory provided by semiconductor circuits.

serial A sequential mode of operation of bits.

serial data transmission The sequential transmission of data over the same wire or bus.

serial highway A highway (or bus) handling data transmission serially.

simplex The transmission only or the reception only of data.

SMR Series Mode Rejection – the rejection of fluctuations superimposed in series with a signal.

software A set of programs, subroutines, executive, compiler, etc, for the operation of a computer system.

source A program written in other than machine code.

SSI Small-Scale Integration (circuit).

subroutine A sequence of instructions to carry out a repetitive task within a program.

synchronization The requirement that several devices or instruments operate at the same instant, controlled from a common command.

synchronous transfer The simultaneous transmission of data between devices by use of a common timing (clock) signal.

talker A device which, when selectively addressed, can generate and transmit data in the IEC 625 interface system. Only one talker can function at any one time.

teletype (TTY) A slow peripheral comprised of a transmitting keyboard and a receiver/printer.

terminal An input/output device such as a teletype or VDU.

transfer rate The speed of data transmission between devices in bits, or words/sec, etc.

translator A program to convert from a source language to machine code.

TTL Transistor Transistor Logic – biopolar manufacturing technique.

TTY *See* teletype.

UART A Universal Asynchronous Receiver and Transmitter used in serial data transmission.

uniline message an interface message sent out on one signal line only in the IEC 625 interface system.

Unibus The bus used in PDP11 series computer for I/O and memory transfers.

utility A standard program used for facilitate program preparation, disk handling, etc.

USART A Universal Synchronous and Asynchronous Receiver and Transmitter used in serial data transmission.

USRT A Universal Synchronous Receiver and Transmitter used in serial data transmission.

VDU Visual Display Unit – a peripheral device displaying alphanumeric data on a cathode-ray tube screen.

VLSI Very-Large-Scale Integration (circuit).

VSWR Voltage Standing Wave Ratio – a measure of circuit imperfections.

V24 A bit serial data transmission standard established by the CCITT

watchdog timer A device to monitor correct functioning of CPU and devices. It causes an alarm function if a machine code instruction is not completed in a specified time.

word A group of bits used in a system, usually in multiples of 8.

word length The number of bits that can be written to or read from memory during a single cycle; greater word length corresponds to greater accuracy.

Appendices

Characteristics of A/D and D/A converters

Number of bits n	2^n	Resolution %	ppm
0	1	100.0	1000000
1	2	50.0	500000
2	4	25.0	250000
3	8	12.5	125000
4	16	6.25	62500
5	32	3.125	15625
6	64	1.563	31250
7	128	0.781	7812
8	256	0.391	3906
9	512	0.195	1953
10	1024	0.0977	977
11	2048	0.0488	488
12	4096	0.0244	244
13	8192	0.0122	122
14	16384	0.0061	61
15	32768	0.00305	31
16	65536	0.00153	15

Relation of voltage proportions expressed in decibels and ratios

$\frac{U_2}{U_1}$, dB	$\frac{U_2}{U_1}$	$\frac{U_2}{U_1}$, dB	$\frac{U_2}{U_1}$
0	1.000	15	5.623
1	1.122	16	6.309
2	1.259	17	7.079
3	1.413	18	7.943
4	1.585	19	8.913
5	1.778	20	10
6	1.995	40	10^2
7	2.239	60	10^3
8	2.512	80	10^4
9	2.818	100	10^5
10	3.162	120	10^6
11	3.548	140	10^7
12	3.981	160	10^8
13	4.467	180	10^9
14	5.012	200	10^{10}

Octal/hexadecimal/ASCII code tables

Octal	*Hexa-decimal*	*ASCII*	*Octal*	*Hexa-decimal*	*ASCII*
000	00	NUL	100	40	@
001	01	SOH	101	41	A
002	02	STX	102	42	B
003	03	ETX	103	43	C
004	04	EOT	104	44	D
005	05	ENQ	105	45	E
006	06	ACK	106	46	F
007	07	BEL	107	47	G
010	08	BS	110	48	H
011	09	HT	111	49	I
012	0A	LF	112	4A	J
013	0B	VT	113	4B	K
014	0C	FF	114	4C	L
015	0D	CR	115	4D	M
016	0E	SO	116	4E	N
017	0F	SI	117	4F	O
020	10	DLE	120	50	P
021	11	DC1	121	51	Q
022	12	DC2	122	52	R
023	13	DC3	123	53	S
024	14	DC4	124	54	T
025	15	NAK	125	55	U
026	16	SYN	126	56	V
027	17	ETB	127	57	W
030	18	CAN	130	58	X
031	19	EM	131	59	Y
032	1A	SUB	132	5A	Z
033	1B	ESC	133	5B	[
034	1C	FS	134	5C	\
035	1D	GS	135	5D	]
036	1E	RS	136	5E	Λ
037	1F	US	137	5F	–
040	20	SP	140	60	`
041	21	!	141	61	a
042	22	“	142	62	b
043	23	#	143	63	c
044	24	$	144	64	d
045	25	%	145	65	e
046	26	&	146	66	f
047	27	'	147	67	g
050	28	(	150	68	h
051	29	)	151	69	i
052	2A	*	152	6A	j
053	2B	+	153	6B	k
054	2C	,	154	6C	l
055	2D	–	155	6D	m
056	2E	.	156	6E	n
057	2F	/	157	6F	o
060	30	0	160	70	p
061	31	1	161	71	q
062	32	2	162	72	r
063	33	3	163	73	s
064	34	4	164	74	t
065	35	5	165	75	u
066	36	6	166	76	v
067	37	7	167	77	w
070	38	8	170	78	x
071	39	9	171	79	y
072	3A	:	172	7A	z
073	3B	;	173	7B	{
074	3C	<	174	7C	}
075	3D	=	175	7D	¦
076	3E	>	176	7E	~
077	3F	?	177	7F	DEL

Bibliography

Chapter 1

1 *Communications Interface Primer*, Part 1: *Instruments and Control Systems,* Mar 1978, pp 43–48
2 EIA Standard RS-232-C, *Interface between data terminal equipment employing serial binary data interchange,* IEA, Washington, Aug 1969
3 HAMPEL, A. and ZIMÁNYI, I., 'Korszerű műszerfelépítés – korszerű interface' ['Modern instrument structure – modern interface'], *Mérés és Automatika,* Vol XXI, No 8, 1973, pp 319–323
4 HETTÉSY, J., 'A CAMAC rendszer általános rendszertechnikai felépítése' ['The general system technology structure of the CAMAC system'], *Mérés és Automatika*, Vol XX, No 5, 1973, pp 167–171
5 KNOWLES, R., *Automatic Testing,* McGraw-Hill, New York, 1976
6 LOUGHRY, D. C., 'A new instrument interface: needs and progress toward a standard', *ISA Transactions*, Vol 14(3), 1975, pp 225–230
7 SEBESTYÉN, DR B., *Számítógép-irányítású mérőrendszerek* [*Computer-controlled measuring systems*], Műszaki Könyvkiadó, Budapest, 1976
8 ZIMÁNYI, I., 'Mérésautomatizálási célra alkalmas műszerek interface kialakításai a SIAK és BSI ajánlások tükrében' ['Interface-realizations of instruments suitable for measuring automation purposes with reference to the SIAK and BSI proposals'], *Mérés és Automatika,* Vol XX, No 8, 1972, pp 312–316
9 CAMAC Instrumentation and Interface Standards 1982 (IEEE Standards 583, 595, 596, 675 and 683), IEEE Inc – J. Wiley, New York
10 CAMAC – Updated specifications, (EUR 8500), Vol 1 (1983) and Vol 2 (1984). Commission of the European Communities, Luxemburg
11 IEC document 625-1 and 625-2, *Interface System for Programmable Measuring Apparatus (byte-serial, bit-parallel)* Sales Dept of IEC, 3 Rue Varembe, CH–1211, Geneva 20, Switzerland

Chapter 2

1 CONNEL, R. J., 'Log on to the IEEE interface', *Systems International*, Vol 7, No 2, Feb 1979, pp 31–32
2 IEEE Standard 488–1975, *Digital Interface for Programmable Instrumentation*, The IEEE Inc, New York, Apr 1975
3 *Interface System for Programmable Measuring Apparatus (byte-serial, bit-parallel),* IEC Technical Committee No 66, *Electronic Measuring Equipment,* June 1975
4 LOUGHRY, D. C. and ALLEN, M. S. 'IEEE Standard 488 and microprocessor synergism', *Proceedings of the IEEE,* Vol 66, No 2, Feb 1978, pp 162–172
5 Műszaki Irányelvek [Technical Directives] MI 12049/1–5: *Programmozható elektronikus mérőkészülékek* [*Programmable electronic measuring devices*], Magyar Szabványügyi Hivatal (Hungarian Standards Office), 1976–1979
6 REID, G. and MYLES, R., 'Digital interface bus – the key to network management', *Communications International,* Jan 1979, pp 26–30
7 RICCI, D. W. and STONE, P. S., 'Putting together instrumentation systems at minimum cost', *Hewlett-Packard Journal*, Jan 1975, pp 5–11
8 ZITZMAN, R., 'The IEEE-488 bus, a cost effective solution to system design', WESCON 78 Conf, Session 35, 1978
9 BSI Sales Dept, Linford Wood, Milton Keynes MK14 6LE, UK

Chapter 3

1 ANSI Standard X3 60–1978, *American National Standard for Minimal Basic,* ANSI Inc, New York, Jan 1978
2 ANSI Standard X3 9–1978, *Programming Language FORTRAN,* ANSI Inc, New York, Apr 1978
3 BURTON, D. P. and DEXTER, A. L., *Microprocessor Systems Handbook,* Analog Devices Inc, Norwood, 1977
4 CSÁKÁNY, A. and VAJDA, DR F., *Mikroszámítógépek* [*Microcomputers*], Műszaki Könyvkiadó, Budapest, 1977
5 HOFFMAN, A. J., FRENCH, R. L. and LANG, G. M., 'Minicomputer interfaces: know more, save more', *IEEE Spectrum*, Feb 1974, pp 64–68
6 LESEA, A. and ZAKS, R., *Microprocessor Interfacing Techniques*, Sybex Inc, Berkeley, 1977
7 OSBORNE, A., *An Introduction to Microcomputers*, Osborne & Associates, Berkeley, 1976
8 *PDP-11 Peripherals and Interfacing Handbook*, Digital Equipment Corporation
9 *Performance Evaluation of HP-IB using RTE Operating Systems*, Hewlett-Packard Application Note 201–204
10 SAWIN III, D. H., *Microprocessors and Microcomputer Systems*, Lexington Books, Lexington, 1977
11 ZAKS, R., *Microprocessors from Chips to Systems*, Sybex Inc, Berkeley, 1977

Chapter 4

1 ALLAN, R., 'Instruments and test equipment', *IEEE Spectrum,* Jan 1975, pp 90–94
2 RIBBERO, R. J., *Microprocessors in Instruments and Control,* Wiley, New York, 1977
3 BRUBAKER, R. H. and KLAISS, D. E., 'Microprocessor intelligence in instrumentation simplifies computer software', ISA 77 Conf, NF 77–561
4 DACK, D. G., 'Making the most of microprocessor control', *Hewlett-Packard Journal,* Jan 1976, pp 16–18
5 LEE, R., 'Microprocessor implementation of a measurement instrument and its interface', WESCON 75 Conf, Session 3, 1975
6 LEE, R., 'Microprocessor ICs improve instruments', *Electronic Design,* 26 Apr 1974, pp 150–154
7 NEWMAN, M. and HOOTMAN, J., 'Hardware interfacing techniques for the IEEE bus', WESCON 78 Conf, Session 35, 1978
8 NELSON, J. E., 'Trends in instrumentation', WESCON 76 Conf, Session 24, 1976
9 WIEWALD, J. and WEST, B., 'Recent advances in microprocessor-based test and measuring equipment', WESCON 76 Conf, Session 24, 1976

Chapter 5

1 FRÜHAUF, T., 'Programming automated instrumentation systems for radiotelephones', *Communications International,* Jan 1977, pp 47–53
2 KNOWLES, R., *Automatic Testing,* McGraw-Hill, New York, 1976
3 LŐCS, GY., SARKADI, I. and SZLANKÓ, J., *A BASIC programozási nyelv [The BASIC programming language],* Műszaki Könyvkiadó, Budapest, 1976
4 MAIER, J., 'Conversational RT testing with Tektronix 4051 and SMPU', *News from Rohde & Schwarz,* No 75, 1976, pp 8–10

Chapter 6

1 ABLEIDINGER, B., 'Unravelling the mystery on the GPIB', *Tekscope,* Vol 10, No 2, pp 3–6
2 DIETRICH, H. E., 'Visualising interface bus activity', *Hewlett-Packard Journal,* Jan 1975, pp 19–26
3 FARNBACH, W. A., 'Logic state analysers', *IEEE Transaction on Instrumentation and Measurement,* Dec 1975, pp 353–356
4 KIMBALL, J., 'The IEEE-488 bus – going your way?', *Tekscope, Vol 10, No 2, pp 7–10*
5 *Monitoring the IEEE Bus with the 1602A Logic Stage Analyser,* Hewlett-Packard Application Note 280–2
6 RADNAI, R., *Digitális jelek korszerű vizsgálata és műszerei [Instruments and modern testing of digital signals],* Műszaki Könyvkiadó, Budapest, 1979
7 WIATROWSKI, C. A., 'Monitoring the 488 bus', WESCON 78 Conf, Session 35, 1978

Chapter 7

1 International Telecommunications Union, CCITT: *Proposed New and Revised Series V Recommendations (Document AP VII No 44-E),* General Secretariat, Sales Section, ITU, CCITT, Place des Nations, CH-1211, Geneva 20, Switzerland
2 *The V-Series Report – Standards for Data Transmissions by Telephone,* Bootstrap Ltd, 1981
3 EIA Standards, RS Series, Electronic Industrial Association, Engineering Dept, Standards Sales Office, 2601 Eye Street, NW, Washington DC 20006, USA
4 Hewlett-Packard, Netherland BV, Central Mailing Dept, PO Box 529, 1180 AM, Amstelreen, The Netherlands (various publications)
5 MCNAMARA, J. E., *Technical Aspects of Data Communication,* Digital Equipment Corp, 1978
6 KORN, G. A., *Microprocessors and Small Digital Computer Systems for Engineers and Scientists,* McGraw-Hill, New York, 1978
7 WOOLARD, B. G., *Microprocessors and Microcomputers for Engineering Students and Technicians,* McGraw-Hill, UK, 1981
8 ZAKS, R. and LESEA, A. *Microprocessor Interfacing Techniques,* Sybex Inc, Berkeley, 1979
9 CLULEY, J. C., *Minicomputer and Microprocessor Interfacing,* E. Arnold, 1982
10 *Microprocessors and Microsystems,* Vol 6, No 9, Nov 1982, and Vol 7, No 6, Jly/Aug 1983, Butterworth
11 'Board level design–VME allows for future expansion', *Electronics Industry,* April 1983, E. S. Publications
12 IEEE Specifications – Service Center, 445 Hoes Lane, Piscataway, New Jersey, 08854, USA, or BSI Sales Dept, Linford Wood, Milton Keynes MK14 6LE, UK (various publications)
13 IEEE Proposals, The IEEE Computer Society, Microprocessor Standards Committee, or IEE Working Party on Backplane Buses, The IEE, Savoy Place, London, UK (various publications)
14 Eurobus (ISO DP 6951 BS 6475 : 1984), International Standards Organization, Case Postale 56, CH-1211, Geneva 20, Switzerland, or BSI Sales Dept, Linford Wood, Milton Keynes MK14 6LE, UK
15 G64 – GESPA, SA, 3 Ch des Ulx, CH-1228, Geneva, Plan les Ouates, Switzerland

Index